Springer Series in Computational Physics

W0090687

Editors

J.-L. Armand   H. Cabannes   M. Holt   H. B. Keller
J. Killeen   S. A. Orszag   V. V. Rusanov

Springer Series in Computational Physics

Editors: H. Cabannes, M. Holt, H. B. Keller, J. Killeen, S. A. Orszag, V. V. Rusanov

F. Bauer/ O. Betancourt/ P. Garabedian: A Computational Method in Plasma Physics
1978. vi, 144 pages. 22 figures. ISBN 08833-4

D.L. Book (ed.): Finite-Difference Techniques for Vectorized Fluid Dynamics Calculations
1981. viii, 240 pages. 60 figures. ISBN 10482-8

Y.N. Dnestrovskii/ D.P. Kostomarov: Numerical Simulations of Plasmas
1986. approx. 320 pages. 97 figures. ISBN 15835-9

C.A.J. Fletcher: Computational Galerkin Methods
1984 xi, 309 pages. 107 figures. ISBN 12633-3

R. Glowinski: Numerical Methods for Nonlinear Variational Problems
1984. xv, 493 pages. 80 figures. ISBN 12434-9

R. Gruber/ J. Rappaz: Finite Element Methods in Linear Ideal Magnetohydrodynamics
1985. xi, 180 pages. 103 figures. ISBN 13398-4

M. Holt: Numerical Methods in Fluid Dynamics, 2nd ed.
1984. xi, 273 pages. 114 figures. ISBN 12799-2

J. Killeen/ G.D. Kerbel/ M.C. McCoy/ A.A. Mirin: Computational Methods for Kinetic
Models of Magnetically Confined Plasmas
1986. viii, 240 pages. 77 figures. ISBN 13401-8

M. Kubíček/ M. Marek: Computational Methods in Bifurcation Theory
and Dissipative Structures
1983. xi, 243 pages. 91 figures. ISBN 12070-X

O.G. Mouritsen: Computer Studies of Phase Transitions and Critical Phenomena
1984. xii, 200 pages. 79 figures. ISBN 13397-6

R. Peyret/ T.D. Taylor: Computational Methods for Fluid Flow
1983. x, 358 pages. 129 figures. ISBN 13851-X

O. Pironneau: Optimal Shape Design for Elliptic Systems
1983. xiii, 192 pages. 57 figures. ISBN 12069-6

Yu. I. Shokin: The Method of Differential Approximation
1983. xiii, 296 pages. ISBN 12225-7

D.P. Telionis: Unsteady Viscous Flows
1981. xxiii, 406 pages. 127 figures. ISBN 10481-8

F. Thomasset: Implementation of Finite Element Methods for Navier-Stokes Equations
1981. xii, 161 pages. 86 figures. ISBN 10771-1

J. Killeen   G. D. Kerbel
M. G. McCoy   A. A. Mirin

Computational Methods for Kinetic Models of Magnetically Confined Plasmas

With 77 Illustrations

Springer-Verlag
New York Berlin Heidelberg Tokyo

J. Killeen, G. D. Kerbel, M. G. McCoy, A. A. Mirin

Lawrence Livermore National Laboratory, National MFE Computer Center, University of California, Livermore, CA 94550, U.S.A.

Editors

J.-L. Armand

Department of Mechanical
  Engineering
University of California
Santa Barbara, CA 93106
U.S.A.

H. B. Keller

Applied Mathematics 101-50
Firestone Laboratory
California Institute of
  Technology
Pasadena, CA 91125
U.S.A.

V. V. Rusanov

Keldysh Institute of
  Applied Mathematics
4 Miusskaya pl.
125047 Moscow
U.S.S.R.

Henri Cabannes

Mécanique Théorique
Université Pierre et
  Marie Curie
Tour 66. 4, Place Jussieu
F-75005 Paris
France

John Killeen

Lawrence Livermore
  National Laboratory
P.O. Box 808
Livermore, CA 94550
U.S.A.

M. Holt

Department of Mechanical
  Engineering
College of Engineering
University of California
Berkeley, CA 94720
U.S.A.

Stephen A. Orszag

Department of Mechanical
  and Aerospace Engineering
Princeton, NJ 08544
U.S.A.

Library of Congress Cataloging-in-Publication Data
Main entry under title:
Computational methods for kinetic models of magnetically
  confined plasmas.
  (Springer series in computational physics)
  Bibliography: p.
  Includes index.
  1. Plasma confinement—Mathematical models.
2. Numerical analysis. 3. Tokamaks. I. Killeen, J.
(John), 1925–   II. Title: Kinetic models of
magnetically confined plasmas. III. Series.
QC718.5.C65C65   1986   530.4'4   85–25176

© 1986 by Springer-Verlag New York Inc.
Softcover reprint of the hardcover 1st edition   1986
All rights reserved. No part of this book may be translated or reproduced in any form without written permission from Springer-Verlag, 175 Fifth Avenue, New York, New York 10010, U.S.A.

Typeset by Asco Trade Typesetting Ltd, North Point, Hong Kong.

9 8 7 6 5 4 3 2 1

ISBN 978-3-642-85956-4   ISBN 978-3-642-85954-0 (eBook)
DOI 10.1007/978-3-642-85954-0

Preface

Because magnetically confined plasmas are generally not found in a state of thermodynamic equilibrium, they have been studied extensively with methods of applied kinetic theory. In closed magnetic field line confinement devices such as the tokamak, non-Maxwellian distortions usually occur as a result of auxiliary heating and transport. In magnetic mirror configurations even the intended steady state plasma is far from local thermodynamic equilibrium because of losses along open magnetic field lines. In both of these major fusion devices, kinetic models based on the Boltzmann equation with Fokker–Planck collision terms have been successful in representing plasma behavior. The heating of plasmas by energetic neutral beams or microwaves, the production and thermalization of α-particles in thermonuclear reactor plasmas, the study of runaway electrons in tokamaks, and the performance of two-energy component fusion reactors are some examples of processes in which the solution of kinetic equations is appropriate and, moreover, generally necessary for an understanding of the plasma dynamics.

Ultimately, the problem is to solve a nonlinear partial differential equation for the distribution function of each charged plasma species in terms of six phase space variables and time. The dimensionality of the problem may be reduced through imposing certain symmetry conditions. For example, fewer spatial dimensions are needed if either the magnetic field is taken to be uniform or the magnetic field inhomogeneity enters principally through its variation along the direction of the field. Velocity dimensionality is reduced through phase averaging over the angular coordinates associated with nearly recurrent motion (e.g., gyro-motion or bounce motion). Four independent variables— one spatial coordinate, two velocity space coordinates, and time—are sufficient for many applications. The evolution of numerical Fokker–Planck calculations over the past 25 years has capitalized on the tractability afforded by these basic considerations.

This book describes a number of state-of-the-art Fokker–Planck computational models developed over the past decade, including (i) an optimized uniform-field package easily adapted to a wide variety of problems; (ii) a bounce-averaged model, featuring a quasilinear RF operator, which is applicable to heating and current drive in tokamaks at either the electron cyclotron, ion cyclotron, or lower hybrid frequency; and (iii) a radial Fokker–Planck/

transport code applicable to tokamaks with intense neutral beam heating. The motivation behind the book is to bring together under one cover a unified description of this research, which has been carried out at the National Magnetic Fusion Energy Computer Center at Livermore.

The authors have benefited from earlier collaborations with Warren Heckrotte, Archer Futch, Kenneth Marx and Marvin Rensink at Livermore and more recent interaction with Dan Jassby of Princeton Plasma Physics Laboratory and Robert Harvey of G.A. Technologies. The authors wish to thank Ms Louise Beite for the excellent job of typing the manuscript. Her patience in dealing with four authors has been remarkable.

Livermore   J. Killeen
California 1985   G. D. Kerbel
                                                                    M. G. McCoy
                                                                    A. A. Mirin

Contents

CHAPTER 1

Introduction

In the simulation of magnetically confined plasmas, where the charged particle distribution functions are not Maxwellian and where a knowledge of those distribution functions is important, kinetic equations must be solved. For instance, the proposition that a stable mirror plasma will yield net thermonuclear power depends on the rate at which particles are lost out of the ends of the device. At number densities and energies typical of mirror machines, the end losses are due primarily to the scattering of charged particles by classical Coulomb collisions into regions of velocity space, known as "loss cones", where they are lost almost immediately. The kinetic equation describing this process is the Boltzmann equation with Fokker–Planck collision terms (Rosenbluth *et al.*, 1957).

The use of kinetic equations is not restricted to mirror systems. The heating of plasmas by energetic neutral beams or microwaves, the thermalization of α-particles in D–T plasmas, the study of runaway electrons in tokamaks, and the performance of two-energy component fusion reactors are other examples where the solution of the kinetic equations is required.

The problem is to solve a nonlinear partial differential equation for the distribution function of each charged species in the plasma in terms of seven independent variables (three spatial coordinates, three velocity coordinates, and time). Since such an equation, even for a single species, exceeds the capability of any present computer, simplifications are required to make the problem tractable. The number of spatial dimensions can be reduced in the event that either the magnetic field is uniform (or square well), or as is often the case in tokamaks and mirrors, that the magnetic field inhomogeneity enters the problem principally through its variation along the direction of the field (i.e., the perpendicular drift velocity is neglected relative to the thermal velocity). Further reduction is realized by gyro-phase averaging, thus eliminating from consideration any dependence on the gyro-phase. This is warranted by the large difference between the time scale of the gyro-motion and that of the collisional relaxation of the population of gyro-center orbits. As a consequence, for many applications the number of independent variables may be reduced to four: one spatial variable, z; two velocity space coordinates, v and θ, the speed and pitch angle; and time, t. The increased tractability afforded by these basic assumptions, or more restrictive ones, has fostered an evolution of numerical Fokker–Planck calculations over the past 25 years.

The work of Roberts and Carr (1960) and Bing and Roberts (1961) on mirror machines consisted of a solution of the Fokker–Planck equation for one species of ions only, ignoring the effects of the electrons and spatial inhomogeneities except for the existence of a magnetic loss cone. They also investigated the adequacy of an approximation to the Fokker–Planck equation in which the solution was assumed to be approximately separable in the velocity coordinates v and θ.

BenDaniel and Allis (1962) extended the work of Roberts and Carr, particularly in the area of approximating the actual solution by a separated solution. They also made some progress toward approximate solutions in the case where spatial inhomogeneities exist.

Codes applicable to more than one species were then developed. Killeen *et al.* (1962) studied the energy transfer from hot ions to cold electrons by solving the isotropic electron Fokker–Planck equation, keeping the hot ion distribution fixed in time. Killeen and Futch (1968) and Fowler and Rankin (1962, 1966) solved the Fokker–Planck equations for both ions and electrons, assuming that the evolution of the distribution functions could be described by the equations for isotropic distributions, with certain factors included to take into account the presence of the loss cone and ambipolar potential. The Fowler and Rankin code was for a steady-state model, whereas the Killeen and Futch code was time-dependent and included the effects of charge exchange and build-up of a plasma formed by neutral injection.

A full multispecies model (Killeen and Mirin, 1970, 1978) was developed in order to study beam-driven D–T and D–^{3}He mirror reactors, including the effects of reaction products. In this model the "Rosenbluth potentials" are taken to be isotropic and the distribution functions are represented by their lowest angular eigenfunction. An extensive parameter study (Futch *et al.*, 1972) was conducted, yielding values of the confinement parameter $n\tau$ and the figure of merit Q (the ratio of thermonuclear power to injected power) as a function of mirror ratio and injection energy.

This model is described in detail in Section 2.2, where the numerical methods and several applications are discussed. A modification of this model, to include anisotropic Rosenbluth potentials, is given in Section 2.3.3.

In a Ph.D. thesis, Werkoff (1973) examined the consequences of assuming the Rosenbluth potentials to be isotropic and the distribution functions to be representable by the lowest normal mode. His initial conclusions were that including the anisotropic part of the ion–ion Rosenbluth potentials substantially increased the value of $n\tau$ for mirror containment. A two-dimensional (spectral in θ) numerical model to calculate this effect is described in Section 2.3. The model is a multispecies treatment, as in Section 2.2, and uses an orthogonal series in θ for the distribution functions and a Legendre polynomial expansion in θ for the Rosenbluth potentials.

Development of models using two velocity space coordinates, v and θ, also took place in the 1970s. Killeen and Marx (1970) developed a code which solved the unseparated Fokker–Planck equation in v and θ for a single ion species, under the assumption that the electrons could be represented by a Maxwellian distribution function with loss cone removed. The potential difference between the mirror throat and midplane, which has a substantial effect on the velocity space loss cone, is determined by iteration on the electron equations in such a way as to equalize the positive and negative charge densities.

Mirin then generalized this to a multispecies model (Killeen *et al.*, 1976; Killeen and Mirin, 1978). Various improvements were made to the previous model. The Fokker–Planck operator is numerically differenced so as to conserve density. The Rosenbluth potentials and their derivatives are computed through expansions in Legendre polynomials.

The reasons for the development of a two-dimensional finite-difference code, given the availability of the two-dimensional expansion code of Section 2.3, were speed and consistency. For a narrow angular source, several angular normal modes are needed to represent the distribution function. Comparisons between the expansion code and the single-species, two-dimensional finite-difference code showed the latter to be much faster. Furthermore, as discussed in Section 2.3.1, for mirror problems with a varying potential along a field line, there is a lack of consistency, as the normal modes are computed neglecting that variation in potential; i.e., since the angular extent of the domain is a function of v, the problem is not truly separable. The two-dimensional formulation described above resolves this problem.

In Section 2.4, the standard two-dimensional, multispecies, nonlinear Fokker–Planck code, FPPAC (McCoy *et al.*, 1981) is described. This code has been put in package form and optimized for the Cray-1 vector computer (McCoy *et al.*, 1979), and it has become the standard nonlinear Fokker–Planck code for plasmas in uniform magnetic fields.

However, it is generally the case in magnetic fusion devices that the magnetic field is nonuniform. Motion along the direction of the field carries the particle through finite variations of the field as it executes its motion. It is often necessary to consider this nonuniformity in order to include features distinctive of a particular device.

An example of such a feature is the presence of trapped particles in tokamaks. Gyro-center orbits which are nearly trapped and oppositely directed may be topologically adjacent in a tokamak and thus may populate one another through diffusion (collisional or quasilinear). In a uniform magnetic field calculation this could not occur. The only recurrent motion in that case is the gyro-motion, and orbits oppositely directed in the preceding sense are inaccessible to one another through small angle collisions or other similarly diffusive processes.

The electrical conductivity of a tokamak plasma in the direction of the magnetic field is affected by the presence of a population of trapped electrons (Hinton and Hazeltine, 1976; Conner *et al.*, 1973). Since most trapped particles execute nearly recurrent "bounce" orbits along the direction of the ohmic electric field, the effect of the field on them is effectively nullified, averaged away by their motion. Trapped particles thus carry no electrical current and cannot contribute to the plasma conductivity.

The use of cyclotron resonance heating in fusion devices and the production of plasma currents, to allow steady-state operation by the application of radio frequency (RF) microwave fields, is currently of great interest. The nonuniformity of the magnetic field bears upon each of these phenomena.

The first code to treat realistic spatial dependence of the confining magnetic field (z-dependence) was developed by Marx (Killeen and Marx, 1970; Marx, 1970). This model, applicable to mirror devices, averages the fast bounce motion of the ion gyro-centers (bounce-averaging) and includes the effects of the axial variation of the ambipolar potential.

Cutler (Cutler *et al.*, 1977) subsequently wrote an improved version which bounce-averages the Fokker–Planck coefficients instead of numerically integrating the distribution functions at each value of z. The early bounce-averaged codes were written in nonconservative form and modeled essentially single region problems. Only trapped particle distributions were evolved; passing distributions were fixed.

With the advent of the tandem mirror concept, it became necessary to consider scenarios with more than one magnetic well. This required the numerical implementation of generalized conditions linking solutions on a multisheeted, two-dimensional domain across phase space flux separatrices. A multiregion bounce-average code using the finite-element technique was developed by Matsuda (Matsuda and Stewart, 1985). This formulation also includes a relativistic formulation of the Fokker–Planck operator as well as a quasilinear RF model.

Efforts to develop bounce-averaged Fokker–Planck codes for modeling tokamaks encountered the same difficulty associated with multiwell mirror problems. Tokamaks are closed field line devices, and as such, trapped and circulating distributions coexist on a single field line typically for many bounce times; the problem is inherently multiregional. Presuming axisymmetry reduces the analysis to that of a single well in a periodic well structure. Nevertheless, the problem still requires the same generalized conditions linking solutions across phase-flow separatrices. The tokamak problem, however, benefits from the simplification that the variation of potential along field lines may be neglected, thereby streamlining the representation of gyro-center orbits and the bounce-averaging algorithm.

One of the first bounce-averaged Fokker–Planck codes for tokamaks (Goldston, 1977) was applied to the problem of calculating charge-exchange spectra resulting from neutral beam injection in the PLT tokamak at Princeton. The code is linear and nonconservative, has a fixed two-dimensional velocity grid and is applicable only to ions. The code was modified to conservative form and generalized by the addition of an RF quasilinear operator by Hammett (Hammett *et al.*, 1983).

Kerbel and McCoy (1985) developed a multispecies, bounce-averaged, nonlinear Fokker–Planck code with an RF quasilinear model equally applicable to ions or electrons. Chapter 3, which is devoted to a description of this model, is divided into three topical areas: mathematical development, numerical analysis, and applications. In Sections 3.1.1 and 3.1.2 the partial differential operators modeling the effects of collisional and resonant diffusion are derived and reduced to a form amenable to numerical analysis. Section 3.1.3 is devoted to an examination of the boundary conditions in velocity space. Particular attention is given to the boundary layer separating trapped and circulating orbits to allow meaningful numerical implementation. Source and loss terms are given in Section 3.1.4. Velocity space loss region models are discussed in Section 3.1.5.

The second part of the chapter, in combination with the appendices, is devoted to numerical procedures employed to compute the relevant operator coefficients and to time-advance the equations. The problem of density conservation and the treatment of the boundary layer between trapped and circulating orbit regions in velocity space is discussed.

The last part of the chapter outlines some illustrative examples of the capabilities of the model. Included are a unified calculation of neoclassical corrections to the classical resistivity of a tokamak plasma, for finite aspect ratio, and the simulation of a scanning charge-exchange spectrum analyzer in an ICRF excited discharge.

The modeling of neutral beam injection in tokamaks and the subsequent evolution of the fast ion distribution function requires the solution of Fokker–Planck equations. The first injection of neutral beams into tokamak plasmas took place at the Culham, Princeton, and Oak Rige Laboratories in 1972–73. The injected ions were studied with linearized Fokker–Planck models (Cordey and Core, 1974; Gaffey, 1976; Callen *et al.*, 1975) and the expected plasma heating was observed experimentally.

With the advent of much more powerful neutral beams, it became possible to consider neutral–beam–driven tokamak fusion reactors. For such devices, three operating regimes (Jassby, 1977) can be considered: (1) the beam-driven thermonuclear reactor; (2) the two-energy component torus (TCT); and (3) the energetic ion reactor, e.g., the counterstreaming ion torus (CIT). In order to study reactors in regimes (2) or (3), a nonlinear Fokker–Planck model must be used because most of the fusion energy is produced by beam–beam or beam–plasma reactions. Furthermore, when co- and counterinjection are used, or major radius compression is employed, a two velocity space dimensional Fokker–Planck operator is required.

Fortunately, the nonlinear, two-dimensional, multispecies Fokker–Planck model described in Section 2.4 had been developed for the mirror program in

1973. This model was applied successfully to several scenarios of TCT operation (Killeen *et al.*, 1974; Berk *et al.*, 1975; Killeen *et al.*, 1975).

An important element of these simulations is the calculation of the energy multiplication factor, Q, defined in Section 2.2.2, for the various operating scenarios. This involves an accurate calculation of $\langle \sigma v \rangle$ for each pair of reacting species. The methods developed for computing these multidimensional integrals are reported elsewhere (Marx *et al.*, 1976, Cordey *et al.*, 1978) and are briefly reviewed in Section 4.1.4.

Neutral-beam-heated tokamaks are characterized by the presence of one or more energetic species which are quite non-Maxwellian, along with a warm Maxwellian bulk plasma. This background plasma may be described by a set of fluid equations. However, for cases in which there is a large energetic ion population, it is very important to represent the energetic species by means of velocity space distribution functions and to follow their evolution in time by integrating the Fokker–Planck equations. It is essential to utilize the full nonlinear Fokker–Planck operator to assure that the slowing down and scattering of these energetic species are computed accurately and realistically.

The successful application of the two-dimensional Fokker–Planck model to the energy multiplication studies of TCT led to the formulation of a more complete model of beam-driven tokamak behavior (Mirin *et al.*, 1977; Killeen *et al.*, 1981). The Fokker–Planck/Transport (FPT) code in its present form is described in Section 4.1.

This model, in addition to solving one-dimensional radial transport equations for the bulk plasma densities and temperatures, solves *nonlinear* Fokker–Planck equations in two-dimensional velocity space for the energetic ion distribution functions, as a function of minor radius in the tokamak. This formulation contrasts with that of Fowler *et al.* (1978), in which the fast ions are described by a linear treatment. The FPT code has been evolving since 1975, and it has been applied to a CIT reactor study (Jassby *et al.*, 1977), to the large Princeton tokamaks (Sections 4.2.1 and 4.2.2), and to the Culham DITE tokamak (Section 4.2.3).

References

D. J. BenDaniel and W. P. Allis, *Plasma Phys.*, **4**, 31 (1962).

H. L. Berk, H. P. Furth, D. L. Jassby, R. M. Kulsrud, C. S. Liu, M. N. Rosenbluth, P. H. Rutherford, F. H. Tenney, T. Johnson, J. Killeen, A. A. Mirin, M. E. Rensink, and C. W. Horton, Jr., Plasma Physics and Controlled Nuclear Fusion Research, 1974 (IAEA, Vienna) III, 569 (1975).

G. Bing and J. E. Roberts, *Phys. Fluids*, **4**, 1039 (1961).

J. D. Callen *et al.*, Plasma Physics and Controlled Nuclear Fusion Research, 1974 (IAEA, Vienna), I, 645 (1975).

J. W. Connor, R. C. Grimm, R. J. Hastie, and P. M. Keeping, *Nucl. Fusion*, **13**, 211 (1973).

J. G. Cordey and W. G. F. Core, *Phys. Fluids*, **17**, 1626 (1974).

J. G. Cordey, K. D. Marx, M. G. McCoy, A. A. Mirin, and M. E. Rensink, *J. Comput. Phys.*, **28**, 115 (1978).

T. A. Cutler, L. D. Pearlstein, and M. E. Rensink, "Computation of the Bounce-Average Code," Report UCRL-52233, Lawrence Livermore National Laboratory, Livermore, CA (1977).

R. H. Fowler, J. Smith, and J. A. Rome, *Comput. Phys. Commun.*, **13**, 323 (1978).

T. K. Fowler and M. Rankin, *Plasma Phys.*, **4**, 311 (1962).

T. K. Fowler and M. Rankin, *Plasma Phys.*, **8**, 121 (1966).

A. H. Futch, Jr., J. P. Holdren, J. Killeen, and A. A. Mirin, *Plasma Phys.*, **14**, 211 (1972).

J. D. Gaffey, *J. Plasma Phys.*, **16**, 149 (1976).

R. J. Goldston, Ph.D. Thesis, Princeton University, Princeton, NJ (1977).

G. W. Hammett, J. C. Hosea, R. J. Goldston, D. Q. Hwang, R. Kaita, D. M. Manos, and J.R. Wilson, "Fast ion charge-exchange measurements during ICRF heating in PLT," *Bull. Amer. Phys. Soc.* **28**, 1129 APS Meeting, Los Angeles (1983).

F. L. Hinton and R. D. Hazeltine, *Phys. Fluids*, **48**, 239 (1976).

D. L. Jassby, *Nucl. Fusion*, **17**, 3009 (1977).

D. L. Jassby, R. M. Kulsrud, F. W. Perkins, J. Killeen, K. D. Marx, M. G. McCoy, A. A. Mirin, M. E. Rensink, and C. G. Tull, Plasma Physics and Controlled Nuclear Fusion Research, 1976 (IAEA, Vienna) II, 435 (1977).

G. D. Kerbel and M. G. McCoy, "Kinetic Theory and Simulation of Multispecies Plasmas in Tokamaks Excited with ICRF Microwaves," Report UCRL-92062, Lawrence Livermore National Laboratory, Livermore, CA (1985). To appear in *Phys. Fluids*.

J. Killeen and A. H. Futch, Jr., *J. Comput. Phys.*, **2**, 236 (1968).

J. Killeen, W. Heckrotte, and G. Boer, *Nucl. Fusion*, Part 1, 183 (1962).

J. Killeen, T. H. Johnson, A. A. Mirin, and M. E. Rensink, "Computational Studies of the Two-Component Toroidal Fusion Test Reactor," Report UCID-16530, Lawrence Livermore National Laboratory, Livermore, CA (1974).

J. Killeen and K. D. Marx, *Meth. Comput. Phys.*, **9**, 422 (1970).

J. Killeen, K. D. Marx, A. A. Mirin, and M. E. Rensink, Proceedings of the Seventh European Conference on Controlled Fusion and Plasma Physics, Lausanne, 1975, **1**, 22 (1975).

J. Killeen and A. A. Mirin, Fourth Conference on Numerical Simulation of Plasmas, NRL, Washington, DC, 685 (1970).

J. Killeen, A. A. Mirin, and M. E. Rensink, *Meth. Comput. Phys.*, **16**, 389 (1976).

J. Killeen and A. A. Mirin, College in Theoretical and Computational Plasma Physics, Trieste (IAEA, Vienna), 27 (1978).

J. Killeen, A. A. Mirin, and M. G. McCoy, Modern Plasma Physics (IAEA, Vienna), 395 (1981).

K. D. Marx, *Phys. Fluids*, **13**, 1355 (1970).

K. D. Marx, A. A. Mirin, M. G. McCoy, M. E. Rensink, and J. Killeen, *Nucl. Fusion*, **16**, 702 (1976).

Y. Matsuda and J. J. Stewart, "A Relativistic Multiregion Bounce-Averaged Fokker–Planck Code for Mirror Plasmas," Report UCRL-92313, Lawrence Livermore National Laboratory, Livermore, CA (1985). Submitted to *J. Comput. Phys.*

M. G. McCoy, A. A. Mirin and J. Killeen, "A Vectorized Fokker–Planck Package for the Cray-1," Scientific Computer Exchange Meeting, Report UCRL-83206, Livermore, CA (1979).

M. G. McCoy, A. A. Mirin, and J. Killeen, *Comput. Phys. Commun.*, **24**, 37 (1981).

A. A. Mirin, J. Killeen, K. D. Marx, and M. E. Rensink, *J. Comput. Phys.*, **23**, 23 (1977).

J. E. Roberts and M. L. Carr, "End-Losses from Mirror Machines," Report UCRL-5651, Lawrence Livermore National Laboratory, Livermore, CA (1960).

M. N. Rosenbluth, W. M. MacDonald, and D. L. Judd, *Phys. Rev.*, **107**, 1 (1957).

F. Werkoff, "Equations de Fokker–Planck avec des Coefficients Anisotropes et Bilan Energéti-que d'un Réacteur á Miroirs," Association Euratom–CEA, Grenoble, France (1973).

CHAPTER 2

Fokker–Planck Models of Multispecies Plasmas in Uniform Magnetic Fields

2.1. Mathematical Model

In this section we consider the collisional kinetic equations as derived in the paper of Rosenbluth *et al.* (1957). We cast these velocity space equations in a form suitable for numerical solution and also consider additional terms representing time-varying forces, sources, and losses. We consider boundary conditions for a full velocity space and a loss cone domain in velocity space.

2.1.1. Fokker–Planck equations

The appropriate kinetic equations are Boltzmann equations with Fokker–Planck collision terms, often referred to simply as Fokker–Planck equations:

$$\frac{\partial f_a}{\partial t} + \mathbf{v} \cdot \frac{\partial f_a}{\partial \mathbf{r}} + \frac{\mathbf{F}}{m_a} \cdot \frac{\partial f_a}{\partial \mathbf{v}} = \left(\frac{\partial f_a}{\partial t} \right)_c + S_a + L_a. \tag{2.1.1}$$

Here f_a is the distribution function in six-dimensional phase space for particles of species a, S_a is a source term, $(\partial f_a / \partial t)_c$ is the collision term, and L_a contains loss terms.

The Fokker–Planck collision term for an inverse-square force was derived by Rosenbluth *et al.* (1957) in the form

$$\frac{1}{\Gamma_a} \left(\frac{\partial f_a}{\partial t} \right)_c = -\frac{\partial}{\partial v_i} \left(f_a \frac{\partial h_a}{\partial v_i} \right) + \frac{1}{2} \frac{\partial^2}{\partial v_i \, \partial v_j} \left(f_a \frac{\partial^2 g_a}{\partial v_i \, \partial v_j} \right), \tag{2.1.2}$$

where $\Gamma_a = 4\pi Z_a^4 e^4 / m_a^2$. In the present work we write the "Rosenbluth potentials"

$$g_a = \sum_b \left(\frac{Z_b}{Z_a} \right)^2 \ln \Lambda_{ab} \int f_b(\mathbf{v}') |\mathbf{v} - \mathbf{v}'| \, d\mathbf{v}', \tag{2.1.3}$$

$$h_a = \sum_b \frac{m_a + m_b}{m_b} \left(\frac{Z_b}{Z_a} \right)^2 \ln \Lambda_{ab} \int f_b(\mathbf{v}') |\mathbf{v} - \mathbf{v}'|^{-1} \, d\mathbf{v}', \tag{2.1.4}$$

which differ from those given by Rosenbluth *et al.* (1957) in the dependence of the Coulomb logarithm on both interacting species and its consequent inclusion under the summations. The expression we use is

$$\ln \Lambda_{ab} = \ln \left\{ \left(\frac{m_a m_b}{m_a + m_b} \right) \frac{2\alpha c \lambda_D}{e^2} \max \left[\left(\frac{2\bar{E}}{m} \right)^{1/2} \right]_{a,b} \right\} - \frac{1}{2}, \tag{2.1.5}$$

where α is the fine structure constant, λ_D the Debye length, $\bar{E}$ the mean energy of particles of species a or b, and the other symbols have their usual meanings. The normalization of f_a in (2.1.1) through (2.1.4) is such that the particle density is given by

$$n_a = \int f_a(\mathbf{v}) \, d\mathbf{v}. \tag{2.1.6}$$

Since the collision term will be seen to contain velocity derivatives of f_a multiplied by velocity moments over f_a, (2.1.1) is a nonlinear, partial, integro-differential equation in seven independent variables. We choose a spherical coordinate system for velocity space (with $\theta = 0$ corresponding to the direction along a magnetic field line) and a cylindrical coordinate system for physical space (z along the magnetic axis). With these coordinate systems, the following assumptions are made:

(1) The system is radially and azimuthally uniform in physical space, and hence, it is also azimuthally symmetric in velocity space. Equivalently, we neglect all gradients transverse to the magnetic field.
(2) The system is axially uniform in physical space. For mirror systems this is equivalent to a magnetic square-well model; for toroidal systems this implies axisymmetry about the major axis of the device.

The transformation of (2.1.2) to spherical polar coordinates (v, θ, ϕ) in velocity space has been given by Rosenbluth et al. (1957). With our assumption of azimuthal symmetry, the resulting distribution functions are of the form $f_a(v, \mu, t)$, where $\mu = \cos \theta$ and $v = |\mathbf{v}|$. The equation for each species is

$$\frac{1}{\Gamma_a} \left(\frac{\partial f_a}{\partial t} \right) = -\frac{1}{v^2} \frac{\partial}{\partial v} \left(f_a v^2 \frac{\partial h_a}{\partial v} \right) - \frac{1}{v^2} \frac{\partial}{\partial \mu} \left[f_a (1 - \mu^2) \frac{\partial h_a}{\partial \mu} \right] + \frac{1}{2v^2} \frac{\partial^2}{\partial v^2} \left(f_a v^2 \frac{\partial^2 g_a}{\partial v^2} \right)$$

$$+ \frac{1}{2v^2} \frac{\partial^2}{\partial \mu^2} \left\{ f_a \left[\frac{1}{v^2} (1 - \mu^2)^2 \frac{\partial^2 g_a}{\partial \mu^2} + \frac{1}{v} (1 - \mu^2) \frac{\partial g_a}{\partial v} \right. \right.$$

$$\left. \left. - \frac{1}{v^2} \mu (1 - \mu^2) \frac{\partial g_a}{\partial \mu} \right] \right\} + \frac{1}{v^2} \frac{\partial^2}{\partial \mu \, \partial v} \left\{ f_a (1 - \mu^2) \left[\frac{\partial^2 g_a}{\partial \mu \, \partial v} - \frac{1}{v} \frac{\partial g_a}{\partial \mu} \right] \right\}$$

$$+ \frac{1}{2v^2} \frac{\partial}{\partial v} \left\{ f_a \left[-\frac{1}{v} (1 - \mu^2) \frac{\partial^2 g_a}{\partial \mu^2} - 2 \frac{\partial g_a}{\partial v} + \frac{2\mu}{v} \frac{\partial g_a}{\partial \mu} \right] \right\} \tag{2.1.7}$$

$$+ \frac{1}{2v^2} \frac{\partial}{\partial \mu} \left\{ f_a \left[\frac{1}{v^2} \mu (1 - \mu^2) \frac{\partial^2 g_a}{\partial \mu^2} + \frac{2\mu}{v} \frac{\partial g_a}{\partial v} \right. \right.$$

$$\left. \left. + \frac{2}{v} (1 - \mu^2) \frac{\partial^2 g_a}{\partial \mu \, \partial v} - \frac{2}{v^2} \frac{\partial g_a}{\partial \mu} \right] \right\}.$$

The functions g_a and h_a, defined by (2.1.3) and (2.1.4), can be represented by expansion in Legendre polynomials (Rosenbluth *et al.*, 1957). For this purpose we let

$$f_a(v, \mu, t) = \sum_{j=0}^{\infty} V_j^a(v, t) P_j(\mu),$$ (2.1.8)

where

$$V_j^a(v, t) = \frac{2j + 1}{2} \int_{-1}^{+1} f_a(v, \mu, t) P_j(\mu) \, d\mu.$$ (2.1.9)

The expansions for g_a and h_a are

$$g_a(v, \mu, t) = \sum_{j=0}^{\infty} \sum_b \left(\frac{Z_b}{Z_a}\right)^2 \ln \Lambda_{ab} B_j^b(v, t) P_j(\mu),$$ (2.1.10)

$$h_a(v, \mu, t) = \sum_{j=0}^{\infty} \sum_b \frac{m_a + m_b}{m_b} \left(\frac{Z_b}{Z_a}\right)^2 \ln \Lambda_{ab} A_j^b(v, t) P_j(\mu),$$ (2.1.11)

where

$$A_j^a = \frac{4\pi}{2j + 1} \left[\int_0^v \frac{(v')^{j+2}}{v^{j+1}} V_j^a(v', t) \, dv' + \int_v^{\infty} \frac{v^j}{(v')^{j-1}} V_j^a(v', t) \, dv' \right],$$ (2.1.12)

$$B_j^a = -\frac{4\pi}{4j^2 - 1} \left[\int_0^v \frac{(v')^{j+2}}{v^{j-1}} \left(1 - \frac{j - \frac{1}{2}}{j + \frac{3}{2}} \frac{(v')^2}{v^2}\right) V_j^a(v') \, dv' \right.$$

$$\left. + \int_v^{\infty} \frac{v^j}{(v')^{j-3}} \left(1 + \frac{j - \frac{1}{2}}{j + \frac{3}{2}} \frac{v^2}{(v')^2}\right) V_j^a(v') \, dv' \right].$$ (2.1.13)

In the computations we take a finite number (which can be varied) of terms in the Legendre expansion of g_a and h_a.

In Section 2.3 we shall describe the solution of (2.1.7) using an expansion in angular eigenfunctions $M_l^a(\mu)$.

In Section 2.4 we describe the finite-difference solution in a two-dimensional velocity space. For this purpose we find it more convenient in differencing and applying boundary conditions to use (v, θ) coordinates rather than (v, μ) coordinates. Equation (2.1.2) in (v, θ) coordinates, written in conservation form, is

$$\frac{1}{\Gamma_a} \left(\frac{\partial f_a}{\partial t}\right)_c = \frac{1}{v^2} \frac{\partial G_a}{\partial v} + \frac{1}{v^2 \sin \theta} \frac{\partial H_a}{\partial \theta},$$ (2.1.14)

where

$$G_a = A_a f_a + B_a \frac{\partial f_a}{\partial v} + C_a \frac{\partial f_a}{\partial \theta},$$ (2.1.15)

$$H_a = D_a f_a + E_a \frac{\partial f_a}{\partial v} + F_a \frac{\partial f_a}{\partial \theta}.$$ (2.1.16)

The coefficients A_a, B_a, C_a, D_a, E_a, and F_a are given by

$$A_a = \frac{v^2}{2} \frac{\partial^3 g_a}{\partial v^3} + v \frac{\partial^2 g_a}{\partial v^2} - \frac{\partial g_a}{\partial v} - v^2 \frac{\partial h_a}{\partial v}$$

$$- \frac{1}{v} \frac{\partial^2 g_a}{\partial \theta^2} + \frac{1}{2} \frac{\partial^3 g_a}{\partial v \partial \theta^2} - \frac{\cot \theta}{v} \frac{\partial g_a}{\partial \theta} + \frac{\cot \theta}{2} \frac{\partial^2 g_a}{\partial v \partial \theta},$$

(2.1.17)

$$B_a = \frac{v^2}{2} \frac{\partial^2 g_a}{\partial v^2},$$

(2.1.18)

$$C_a = - \frac{1}{2v} \frac{\partial g_a}{\partial \theta} + \frac{1}{2} \frac{\partial^2 g_a}{\partial v \partial \theta},$$

(2.1.19)

$$D_a = \frac{\sin \theta}{2v^2} \frac{\partial^3 g_a}{\partial \theta^3} + \frac{\sin \theta}{2} \frac{\partial^3 g_a}{\partial v^2 \partial \theta} + \frac{\sin \theta}{v} \frac{\partial^2 g_a}{\partial v \partial \theta}$$

$$- \frac{1}{2v^2 \sin \theta} \frac{\partial g_a}{\partial \theta} + \frac{\cos \theta}{2v^2} \frac{\partial^2 g_a}{\partial \theta^2} - \sin \theta \frac{\partial h_a}{\partial \theta},$$

(2.1.20)

$$E_a = \sin \theta \left[- \frac{1}{2v} \frac{\partial g_a}{\partial \theta} + \frac{1}{2} \frac{\partial^2 g_a}{\partial v \partial \theta} \right],$$

(2.1.21)

$$F_a = \frac{\sin \theta}{2v^2} \frac{\partial^2 g_a}{\partial \theta^2} + \frac{\sin \theta}{2v} \frac{\partial g_a}{\partial v}.$$

(2.1.22)

In this case the density and energy density of species "a" are defined according to

$$n_a = 2\pi \int \int f_a(v, \theta) v^2 \sin \theta \, dv \, d\theta,$$

(2.1.23)

$$n_a E_a = \pi m_a \int \int f_a(v, \theta) v^4 \sin \theta \, dv \, d\theta.$$

(2.1.24)

Evaluation of the coefficients defined in (2.1.17)–(2.1.22) requires derivatives of the "Rosenbluth potentials". These may be obtained through term-by-term differentiation of (2.1.10)–(2.1.11). To obtain derivatives with respect to v requires differentiation of the right side of (2.1.12)–(2.1.13). This is done analytically. It is convenient to define four functionals

$$M_l(w)(v) = \int_v^\infty w(y) y^{(1-l)} \, dy,$$

(2.1.25)

$$N_l(w)(v) = \int_0^v w(y) y^{(2+l)} \, dy,$$

(2.1.26)

$$R_l(w)(v) = \int_v^\infty w(y) y^{(3-l)} \, dy,$$

(2.1.27)

$$E_l(w)(v) = \int_0^v w(y) y^{(4+l)} \, dy.$$

(2.1.28)

Using these functionals, (2.1.12) and (2.1.13) become

$$A_l^b = \frac{4\pi}{(2l+1)} [v^{-l-1} N_l(V_l^b) + v^l M_l(V_l^b)], \tag{2.1.29}$$

$$B_l^b = \frac{4\pi}{2l+1} \left\{ \frac{1}{(2l+3)} [v^{-l-1} E_l(V_l^b) + v^{l+2} M_l(V_l^b)] \right.$$
$$\left. - \frac{1}{(2l-1)} [v^{1-l} N_l(V_l^b) + v^l R_l(V_l^b)] \right\}. \tag{2.1.30}$$

The resulting expressions are

$$\frac{\partial A_l^b}{\partial v} = \frac{4\pi}{(2l+1)} [lv^{l-1} M_l(V_l^b) - (l+1)v^{-l-2} N_l(V_l^b)], \tag{2.1.31}$$

$$\frac{\partial B_l^b}{\partial v} = \frac{4\pi}{(2l+1)} \left\{ \frac{1}{(2l+3)} [(l+2)v^{l+1} M_l(V_l^b) - (l+1)v^{-l-2} E_l(V_l^b)] \right.$$
$$\left. - \frac{1}{(2l-1)} [lv^{l-1} R_l(V_l^b) - (l-1)v^{-l} N_l(V_l^b)] \right\}, \tag{2.1.32}$$

$$\frac{\partial^2 B_l^b}{\partial v^2} = \frac{4\pi}{(2l+1)} \left\{ \frac{(l+1)(l+2)}{(2l+3)} [v^{-l-3} E_l(V_l^b) + v^l M_l(V_l^b)] \right.$$
$$\left. - \frac{l(l-1)}{(2l-1)} [v^{-l-1} N_l(V_l^b) + v^{l-2} R_l(V_l^b)] \right\}, \tag{2.1.33}$$

$$\frac{\partial^3 B_l^b}{\partial v^3} = \frac{4\pi}{(2l+1)} \left\{ \frac{1}{(2l+3)} [l(l+1)(l+2)v^{l-1} M_l(V_l^b) \right.$$
$$- (l+1)(l+2)(l+3)v^{-l-4} E_l(V_l^b)]$$
$$- \frac{1}{(2l-1)} [l(l-1)(l-2)v^{-l-3} R_l(V_l^b) \tag{2.1.34}$$
$$\left. - (l+1)(l)(l-1)v^{-l-2} N_l(V_l^b)] \right\}.$$

Derivatives with respect to θ are also done analytically.

2.1.2. Time-Varying forces

Within the context of our idealized homogeneous plasma model the form of the kinetic equation with time-varying magnetic field is

$$\frac{df}{dt} = \frac{\partial f}{\partial t} + \dot{v}_\parallel \frac{\partial f}{\partial v_\parallel} + \dot{v}_\perp \frac{\partial f}{\partial v_\perp} = \left(\frac{\partial f}{\partial t} \right)_c + S + L, \tag{2.1.35}$$

where $(\dot{v}_\parallel, \dot{v}_\perp)$ is the cyclotron-averaged inductive acceleration. For slowly varying magnetic fields ($\dot{v}_\parallel$, $\dot{v}_\perp$) can be derived from the adiabatic constants of the particle motion.

2.1.2.1. *Magnetic field compression in a mirror machine*

In linear mirror systems the constants of the particle motion are

$$\frac{mv_\perp^2}{2B} = \text{magnetic moment}, \tag{2.1.36}$$

$$2mv_\parallel L = \text{longitudinal invariant}, \tag{2.1.37}$$

where L is the distance between mirrors. To derive ($\dot{v}_\parallel$, $\dot{v}_\perp$) we take the time derivative of these equations, obtaining

$$\frac{2v_\perp \dot{v}_\perp}{B} - \frac{v_\perp^2 \dot{B}}{B^2} = 0, \tag{2.1.38}$$

$$\dot{v}_\parallel L + v_\parallel \dot{L} = 0, \tag{2.1.39}$$

and subsequently

$$\dot{v}_\perp = \tfrac{1}{2} v_\perp \frac{\dot{B}}{B}, \tag{2.1.40}$$

$$\dot{v}_\parallel = -v_\parallel \frac{\dot{L}}{L}. \tag{2.1.41}$$

When these results for $\dot{v}_\parallel$ and $\dot{v}_\perp$ are inserted in (2.1.35) the kinetic equation becomes

$$\frac{\partial f}{\partial t} + \left(\tfrac{1}{2} v \sin^2 \theta \frac{\partial f}{\partial v} + \tfrac{1}{2} \sin \theta \cos \theta \frac{\partial f}{\partial \theta} \right) \frac{\dot{B}}{B} = \left(\frac{\partial f}{\partial t} \right)_c + S + L, \tag{2.1.42}$$

where we have used

$$v_\perp = v \sin \theta, \tag{2.1.43}$$

$$v_\parallel = v \cos \theta, \tag{2.1.44}$$

and longitudinal compression has been neglected. For isotropic distribution functions, $\partial f / \partial \theta$ is zero, and $\sin^2 \theta$ is replaced by its average value $\langle \sin^2 \theta \rangle = \tfrac{2}{3}$ so the one-dimensional kinetic equation becomes

$$\frac{\partial f}{\partial t} + \left(\tfrac{1}{3} v \frac{\partial f}{\partial v} \right) \frac{\dot{B}}{B} = \left(\frac{\partial f}{\partial t} \right)_c + S + L. \tag{2.1.45}$$

2.1.2.2. *Major radius compression in a tokamak*

In axisymmetric toroidal systems the constants of the particle motion are

$$\frac{mv_\perp^2}{2B} = \text{magnetic moment}, \tag{2.1.46}$$

$$mv_\parallel R = \text{toroidal angular momentum}, \tag{2.1.47}$$

where R is the major radius of the torus. For a time-varying major radius (Furth and Yoshikawa, 1970) we derive $(\dot{v}_\parallel, \dot{v}_\perp)$ by taking the time derivative of these equations, obtaining

$$\frac{2v_\perp \dot{v}_\perp}{B} - \frac{v_\perp^2 \dot{B}}{B^2} = 0, \tag{2.1.48}$$

$$\dot{v}_\parallel R + v_\parallel \dot{R} = 0, \tag{2.1.49}$$

and subsequently

$$\dot{v}_\perp = \tfrac{1}{2} v_\perp \frac{\dot{B}}{B}, \tag{2.1.50}$$

$$\dot{v}_\parallel = -v_\parallel \frac{\dot{R}}{R}. \tag{2.1.51}$$

Since B is essentially just the toroidal field strength, it varies inversely with R, yielding,

$$\frac{\dot{B}}{B} = -\frac{\dot{R}}{R}. \tag{2.1.52}$$

When these results are inserted in (2.1.35) the kinetic equation becomes

$$\frac{\partial f}{\partial t} + \frac{\dot{R}}{R}\left[-(1 - \tfrac{1}{2}\sin^2 \theta)v\frac{\partial f}{\partial v} + \tfrac{1}{2}\sin \theta \cos \theta \frac{\partial f}{\partial \theta} \right] = \left(\frac{\partial f}{\partial t}\right)_c + S + L, \tag{2.1.53}$$

where $(\partial f/\partial t)_c$ is given by (2.1.14). For isotropic distribution functions we have

$$\frac{\partial f}{\partial t} + \left(-\tfrac{2}{3}v\frac{\partial f}{\partial v} \right)\frac{\dot{R}}{R} = \left(\frac{\partial f}{\partial t}\right)_c + S + L. \tag{2.1.54}$$

2.1.3. Initial conditions and boundary conditions in velocity space

Since (2.1.14) describes a parabolic partial differential equation for each plasma species distribution function, it is necessary to specify only $f_a(v, \theta, 0)$ (and not its time derivative) as an initial condition for each species.

2.1.3.1. Boundary conditions for a loss cone domain with ambipolar potential

Since applications include the problem of plasma confinement within systems of magnetic mirrors, in this section we consider the mathematical description of such systems.

For a standard mirror, the loss cone angle is (Spitzer, 1962)

$$\sin^2 \theta_{LC} = 1/R_m, \tag{2.1.55}$$

where $R_m = B_m/B(z)$; B_m is the magnetic field at the mirror, and $B(z)$ is the

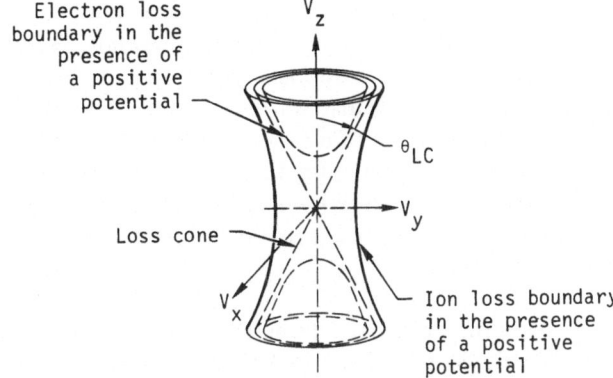

Figure 2.1. Velocity space loss cone boundaries.

magnetic field at the interior point being considered. The orientation of the loss cone in velocity space is displayed in Fig. 2.1. A particle whose angle in velocity space is less than θ_{LC} will be lost from the mirror system. θ_{LC} is independent of velocity as well as particle mass and charge. Equation (2.1.55) is derived under the assumption that no electrostatic potential exists, and θ_{LC} is the actual loss angle only under that condition.

However, because of their greater mobility, the scattering rate of electrons will be greater than that of ions, and more electrons than ions will tend to leak out of the ends of the device. Hence, an ambipolar potential will build up, being greatest at the center and decreasing towards the ends. The fact that this potential is established leads to a fundamental change in the loss characteristics for the two types of particles. The loss regions are then defined by a loss angle which is a function of speed and charge (Kaufman, 1956; Post, 1961; BenDaniel, 1961; Yushmanov, 1966). If $Z_a e$ is charge and ϕ is electrostatic potential, the loss angle is given by

$$\sin^2 \theta_{LC} = 1/R_a, \qquad (2.1.56a)$$

where

$$R_a = \left(\frac{1 + Z_a e\phi/\frac{1}{2}m_a v^2}{R_m} \right)^{-1} \qquad (2.1.56b)$$

is the "effective" mirror ratio.

Equation (2.1.56a) approaches (2.1.55) asymptotically as $v \to \infty$. For ions, the right-hand side of (2.1.56a) can exceed unity; no ion in such a velocity regime can be contained. Conversely, for electrons, this term can be less than zero; all electrons at such velocities will be electrostatically trapped. These regions are shown in Fig. 2.1. The loss region for ions is transformed from a cone into a hyperboloid of one sheet. Its minimum radius occurs at $\theta = \pi/2$, and is equal to the minimum ion velocity possible for confinement. The electron loss region is transformed into a hyperboloid of two sheets.

In the central cell of a tandem mirror (Cohen *et al.*, 1978) the ions and electrons are affected by ambipolar potentials. However, the potential which affects the ions is of the opposite sign and acts to confine them. Hence, the loss region for such ions is also a hyperboloid of two sheets.

In the plug regions of a tandem mirror with thermal barriers the electrons are unconfined magnetically (the mirror ratio is less than 1) but are acted upon by an ambipolar potential which helps confine them. In this case the region of confinement is the *bounded* region defined by

$$\tfrac{1}{2}mv^2 \leq |e\phi|(1 - R_m \sin^2 \theta)^{-1}. \tag{2.1.57}$$

If we assume that particles in the loss cone are lost immediately then we set $f_a = 0$ on the loss cone boundary. If we consider a finite loss time, then we add a loss term described in the next section at meshpoints within the loss cone.

There are certain situations, such as the plug regions of a tandem mirror with thermal barrier, where particles in the loss region are lost locally but not globally; that is, these particles called "passing particles" follow orbits which leave the immediate region of concern but do not leave the machine as a whole. In such a situation the distribution function f_a is set equal to the distribution function of passing particles on the loss boundary.

2.1.3.2. *Boundary conditions for a full velocity space*

In problems where we want to observe the relaxation of a distribution in the absence of a loss cone, or where we assume that ions in a loss cone domain are not lost instantly, full velocity space boundary conditions are applied, namely:

$$f_a(v = 0, \theta) \text{ is independent of } \theta, \tag{2.1.58}$$

$$\frac{\partial f_a}{\partial v}(v = 0, \theta = \pi/2) = 0, \tag{2.1.59}$$

$$f_a(v = \infty, \theta) = 0, \tag{2.1.60}$$

$$\frac{\partial f_a}{\partial \theta}(v, \theta = 0) = \frac{\partial f_a}{\partial \theta}(v, \theta = \pi) = 0. \tag{2.1.61}$$

For problems with symmetry about the angle $\theta = \pi/2$, it is assumed that

$$\frac{\partial f_a}{\partial \theta}(v, \theta = \pi/2) = 0. \tag{2.1.62}$$

2.1.4. Source and loss terms in velocity space

The source term S_a in (2.1.1) is of the form

$$S_a(v, \theta, t) = \sum_l J_a^l(t) S_a(v, \theta) \delta_s^{a,l}(t), \tag{2.1.63}$$

where the shape function $S_a(v, \theta)$ is a Gaussian in v and $\cos \theta$ of density 1, $\delta_s^{a,l}(t)$ is either 0 or 1, and $J_a^l(t)$ is a current of the form

$$J_a^l = \sum_b \left(A_{ab}^l + B_{ab}^l n_b \right). \tag{2.1.64}$$

The quantities A_{ab}^l, and B_{ab}^l are parameters independent of time, and n_b is the density of species "b".

The loss term L_a in (2.1.1) may combine several loss processes. Losses due to charge exchange with the injected neutral beam are expressed as

$$L_a^c = -\left[\sum_{b,l} D_{ab}^l \delta_s^{b,l}(t) \right] f_a(v, \theta, t), \tag{2.1.65}$$

where the quantities D_{ab}^l are constant parameters. The presence of the factors $\delta_s^{b,l}$ allows the whole charge-exchange process to be implemented as a unit; that is, a given charge-exchange term along with its corresponding source term depends on the same temporal function $\delta_s^{b,l}(t)$.

In mirror applications it is often assumed that ions in the loss cone domain are lost instantly. However, it takes a finite length of time for such ions to make one pass between the mirrors before they are actually lost. Thus, the escape time is finite, and for cold ions (as in a target plasma) this time can be quite long. To account for this, we consider a contribution to the loss term L_a of the form

$$L_a^m = -f_a(v, \theta, t) \frac{v_\parallel}{L} \delta^m(v, \theta), \tag{2.1.66}$$

where L is the axial length of the mirror system, and $\delta^m(v, \theta)$ is either 1 or 0, depending on whether or not (v, θ) is in the loss cone domain.

The effects of finite particle and energy confinement times in a tokamak are incorporated by adding a contribution to L_a of the form

$$L_a^\tau = -\frac{f_a(v, \theta, t)}{\tau_p} + \frac{1}{v^2} \frac{\partial}{\partial v} \left[\left(\frac{1}{\tau_e} - \frac{1}{\tau_p} \right) \frac{v^3 f_a(v, \theta, t)}{2} \right], \tag{2.1.67}$$

where τ_p and τ_e are particle and energy confinement times, respectively. If we ignore all other terms in (2.1.1) and compute moments, we find

$$\frac{\partial n_a}{\partial t} = -\frac{n_a}{\tau_p}, \tag{2.1.68}$$

$$\frac{\partial (n_a \bar{E}_a)}{\partial t} = -\frac{n_a \bar{E}_a}{\tau_e}. \tag{2.1.69}$$

The quantity $\bar{E}_a$ is the mean energy of species "a".

2.2. Solution for a Multispecies Plasma in a One-Dimensional Velocity Space

There is a class of problems for which a one-dimensional system of Fokker–Planck equations is useful. Killeen *et al.* (1962) studied the energy transfer from hot ions to cold electrons by solving the isotropic electron Fokker–Planck equation, keeping the hot ion distribution fixed in time. Killeen and Futch (1968) and Fowler and Rankin (1962, 1966) solved the Fokker–Planck equations for both ions and electrons, assuming that the evolution of the distribution functions could be described by the equations for isotropic distributions, with certain factors included to take the presence of the loss cone and ambipolar potential into account. The Fowler and Rankin code was for a steady-state model, whereas the Killeen and Futch code was time-dependent and included the effects of charge-exchange and time-dependent build-up of a plasma formed by neutral injection.

A multispecies model (Killeen and Mirin, 1970, 1978) was developed in order to study beam-driven D–T and D–^{3}He mirror reactors, including the effects of reaction products. The principal assumptions of this model are that the "Rosenbluth potentials" are isotropic and that the distribution functions can be represented by their lowest angular eigenfunction. An extensive parameter study (Futch *et al.*, 1972) was conducted, yielding values of the confinement parameter $n\tau$ and the figure of merit Q (the ratio of thermonuclear power to injected power) as a function of mirror ratio and injection energy.

2.2.1. Numerical methods

In this section we describe the multispecies numerical model which has proved very useful (Futch *et al.*, 1972), in which we assume that the "Rosenbluth potentials" given by (2.1.10) and (2.1.11) are isotropic, i.e.,

$$\frac{\partial g_a}{\partial \mu} = \frac{\partial h_a}{\partial \mu} = 0. \tag{2.2.1}$$

With this assumption (2.1.7) becomes

$$\frac{1}{\Gamma_a}\frac{\partial f_a}{\partial t} = -\frac{1}{v^2}\frac{\partial}{\partial v}\left(f_a v^2 \frac{\partial h_a}{\partial v}\right) + \frac{1}{2v^2}\frac{\partial^2}{\partial v^2}\left(f_a v^2 \frac{\partial^2 g_a}{\partial v^2}\right)$$

$$+ \frac{1}{2v^3}\frac{\partial g_a}{\partial v}\left[(1-\mu^2)\frac{\partial^2 f_a}{\partial \mu^2} - 4\mu\frac{\partial f_a}{\partial \mu} - 2f_a\right] \tag{2.2.2}$$

$$- \frac{1}{v^2}\frac{\partial}{\partial v}\left(f_a\frac{\partial g_a}{\partial v}\right) + \frac{1}{v^3}\frac{\partial g_a}{\partial v}\left(\mu\frac{\partial f_a}{\partial \mu} + f_a\right).$$

Equation (2.2.2) is separable, and if we let

$$f_a(v, \mu, t) = U_a(v, t)M_a(\mu), \tag{2.2.3}$$

then (2.2.2) can be written

$$\frac{M_a}{\Gamma_a}\frac{\partial U_a}{\partial t} = -\frac{M_a}{v^2}\frac{\partial}{\partial v}\left(U_a v^2\frac{\partial h_a}{\partial v}\right) + \frac{M_a}{2v^2}\frac{\partial^2}{\partial v^2}\left(U_a v^2\frac{\partial^2 g_a}{\partial v^2}\right) - \frac{M_a}{v^2}\frac{\partial}{\partial v}\left(U_a\frac{\partial g_a}{\partial v}\right)$$

$$+ \frac{U_a}{2v^3}\frac{\partial g_a}{\partial v}\left[(1 - \mu^2)\frac{\partial^2 M_a}{\partial\mu^2} - 2\mu\frac{\partial M_a}{\partial\mu}\right]. \tag{2.2.4}$$

We obtain an eigenvalue problem for $M_a(\mu)$

$$(1 - \mu^2)\frac{d^2 M_a}{d\mu^2} - 2\mu\frac{dM_a}{d\mu} + \Lambda_a M_a = 0, \tag{2.2.5}$$

which is Legendre's equation. In the full velocity space we have $-1 \le \mu \le 1$; hence the solutions of (2.2.5) are Legendre polynomials, and the lowest mode corresponds to an isotropic distribution function. For a loss cone domain

$$-\cos\theta^a_{LC} < \mu < \cos\theta^a_{LC}, \tag{2.2.6}$$

where θ^a_{LC} is the loss cone angle for species a, defined by (2.1.56). In this case (2.2.5) is solved numerically as an eigenvalue problem with $M_a(\mu) = 0$ at the endpoints. For each eigenvalue Λ_a of (2.2.5) we have an equation of the form [in the following we have replaced $U_a(v, t)$ by $f_a(v, t)$]:

$$\frac{1}{\Gamma_a}\frac{\partial f_a}{\partial t} = -\frac{1}{v^2}\frac{\partial}{\partial v}\left[f_a\left(v^2\frac{\partial h_a}{\partial v} + \frac{\partial g_a}{\partial v}\right)\right] + \frac{1}{2v^2}\frac{\partial^2}{\partial v^2}\left(v^2 f_a\frac{\partial^2 g_a}{\partial v^2}\right)$$

$$-\frac{\Lambda_a}{2v^3}\frac{\partial g_a}{\partial v}f_a. \tag{2.2.7}$$

The last term on the right of (2.2.7) is the particle loss term due to scattering into the loss cone. In this section we consider the lowest eigenvalue, Λ_a.

The functions g_a and h_a defined by (2.1.10) and (2.1.11), in the case where (2.2.1) is assumed, become

$$g_a(v, t) = 4\pi\sum_b\left(\frac{Z_b}{Z_a}\right)^2\ln\Lambda_{ab}\left[\int_0^v f_b(v', t)v\left(1 + \frac{1}{3}\frac{v'^2}{v^2}\right)v'^2\,dv'\right.$$

$$\left. + \int_v^\infty f_b(v', t)\left(1 + \frac{1}{3}\frac{v^2}{v'^2}\right)v'^3\,dv'\right], \tag{2.2.8}$$

$$h_a(v, t) = 4\pi\sum_b\left(\frac{Z_b}{Z_a}\right)^2\frac{m_a + m_b}{m_b}\ln\Lambda_{ab}\left[\int_0^v f_b(v', t)\frac{v'^2}{v}\,dv'\right.$$

$$\left. + \int_v^\infty f_b(v', t)v'\,dv'\right]. \tag{2.2.9}$$

The summations are taken over all the species of particles being considered, including type a. If we use (2.2.8) and (2.2.9) to evaluate the coefficients of (2.2.7), then the equation for $f_a(v, t)$ becomes

$$\frac{\partial f_a}{\partial t} = \frac{1}{v^2} \frac{\partial}{\partial v} \left[\alpha_a f_a + \beta_a \frac{\partial f_a}{\partial v} \right] - \frac{\gamma_a}{v^2} f_a, \tag{2.2.10}$$

where

$$\alpha_a = 4\pi \Gamma_a \sum_b \left[\left(\frac{Z_b}{Z_a} \right)^2 \frac{m_a}{m_b} \ln \Lambda_{ab} \int_0^v f_b(v', t) v'^2 \, dv' \right], \tag{2.2.11}$$

$$\beta_a = 4\pi \Gamma_a \sum_b \left(\frac{Z_b}{Z_a} \right)^2 \ln \Lambda_{ab} \left[\frac{1}{3v} \int_0^v f_b(v', t) v'^4 \, dv' + \frac{v^2}{3} \int_v^\infty f_b(v', t) v' \, dv' \right], \tag{2.2.12}$$

$$\gamma_a = \frac{\Lambda_a}{2v} 4\pi \Gamma_a \sum_b \left(\frac{Z_b}{Z_a} \right)^2 \ln \Lambda_{ab} \left[\int_0^v f_b(v', t) v'^2 \left(1 - \frac{1}{3} \frac{v'^2}{v^2} \right) dv' \right.$$
$$\left. + \frac{2v}{3} \int_v^\infty f_b(v', t) v' \, dv' \right]. \tag{2.2.13}$$

The number density of particles of type a is

$$n_a(t) = 4\pi \int_0^\infty f_a(v, t) v^2 \, dv. \tag{2.2.14}$$

If $\gamma_a = 0$ then $n_a(t)$ should remain constant in time, which gives the appropriate boundary conditions for the solution of (2.2.10), i.e.,

$$\frac{dn_a}{dt} = 4\pi \int_0^\infty \frac{1}{v^2} \frac{\partial}{\partial v} \left[\alpha_a f_a + \beta_a \frac{\partial f_a}{\partial v} \right] v^2 \, dv = 0 \tag{2.2.15}$$

is satisfied by the conditions

$$\left[\alpha_a f_a + \beta_a \frac{\partial f_a}{\partial v} \right] \Bigg|_{v=0}^{v=v_{max}} = 0 \tag{2.2.16}$$

for $t > 0$, where $0 \le v \le v_{max}$ is the domain of (2.2.10). At $v = 0$, (2.2.16) is consistent with $\partial f_a / \partial v = 0$, since $\lim_{v \to 0} \alpha_a(v) / \beta_a(v) = 0$. In multispecies problems with an expelling ambipolar potential, the effective velocity domain will be the interval $[v_{min}, v_{max}]$. In this case v_{min} will depend on the ion species and will equal zero for the electrons.

Equation (2.2.10) is a system of coupled partial differential equations. For each particle species, a, we have an equation corresponding to Λ_a, the eigenvalue of (2.2.5). In the next section we consider several eigenvalues for each species, but in this section we have used only the lowest eigenvalue corresponding to the first normal mode. Hence we have one equation for each species. In solving the system given by (2.2.10) we consider it as a vector equation of the form

$$\frac{\partial F}{\partial t} = \frac{1}{v^2}\frac{\partial G}{\partial v} - \frac{C}{v^2}F + D, \tag{2.2.17}$$

where

$$F = \begin{bmatrix} f_1 \\ \vdots \\ f_P \end{bmatrix}; \qquad G = AF + B\frac{\partial F}{\partial v}, \tag{2.2.18}$$

and A, B, C are diagonal $p \times p$ matrices and D is the source vector. Without source and loss terms, (2.2.17) becomes

$$\frac{\partial F}{\partial t} = \frac{1}{v^2}\frac{\partial G}{\partial v}. \tag{2.2.19}$$

We see that this equation is in conservation form (divergence of a flux), which is consistent with the correct boundary conditions.

We solve (2.2.17) by finite-difference methods. On the domain $0 \le v \le v_J$, $t \ge 0$, we have a finite-difference mesh denoted by $v_j, j = 0, \ldots, J$ and $t_n, n = 0$, 1, 2, ... They v spacing is variable and we define $\Delta v_{j+1/2} = v_{j+1} - v_j$, $\Delta v_{j-1/2} = v_j - v_{j-1}$, $\Delta v_j = \frac{1}{2}(v_{j+1} - v_{j-1})$. We approximate (2.2.17) by the following implicit difference equation:

$$\begin{aligned}
\frac{F_j^{n+1} - F_j^n}{\Delta t} &= \rho\left[\frac{1}{v_j^2}\frac{G_{j+1/2}^{n+1} - G_{j-1/2}^{n+1}}{\Delta v_j} - \frac{C_j^{n+1}}{v_j^2}F_j^{n+1} + D_j^{n+1}\right] \\
&+ (1-\rho)\left[\frac{1}{v_j^2}\frac{G_{j+1/2}^n - G_{j-1/2}^n}{\Delta v_j} - \frac{C_j^n}{v_j^2}F_j^n + D_j^n\right],
\end{aligned} \tag{2.2.20}$$

where

$$G_{j+1/2}^n = \frac{1}{2}A_{j+1/2}^n(F_{j+1}^n + F_j^n) + B_{j+1/2}^n\frac{F_{j+1}^n - F_j^n}{\Delta v_{j+1/2}}, \tag{2.2.21}$$

$$G_{j-1/2}^n = \frac{1}{2}A_{j-1/2}^n(F_j^n + F_{j-1}^n) + B_{j-1/2}^n\frac{F_j^n - F_{j-1}^n}{\Delta v_{j-1/2}}. \tag{2.2.22}$$

For numerical stability we must have $\frac{1}{2} \le \rho \le 1$; we usually take $\rho = 1$. Without source and loss terms, i.e., $C_j^n = D_j^n = 0$ for all j and n, we have

$$\sum_{j=1}^{J-1}\left(\frac{F_j^{n+1} - F_j^n}{\Delta t}\right)v_j^2\Delta v_j = 0 \tag{2.2.23}$$

for all n, independent of the mesh spacing, as long as $G_{1/2}^n = G_{J-1/2}^n = 0$, for all n. This condition is the boundary condition given by (2.2.16). Thus we see that our difference scheme rigorously conserves particle density in the absence of source and loss terms.

In order to solve the difference equations given by (2.2.20), we write it as the linear algebraic system:

$$-\alpha_j^{n+1} F_{j+1}^{n+1} + \beta_j^{n+1} F_j^{n+1} - \gamma_j^{n+1} F_{j-1}^{n+1} = \delta_j^n, \tag{2.2.24}$$

$j = 1, \ldots, J - 1$, where

$$\alpha_j^{n+1} = \frac{\rho}{v_j^2} \frac{1}{\Delta v_j} \left(\frac{A_{j+1/2}^{n+1}}{2} + \frac{B_{j+1/2}^{n+1}}{\Delta v_{j+1/2}} \right), \tag{2.2.25}$$

$$\beta_j^{n+1} = \frac{1}{\Delta t} - \frac{\rho}{v_j^2} \left[\frac{1}{\Delta v_j} \left(\frac{A_{j+1/2}^{n+1} - A_{j-1/2}^{n+1}}{2} - \frac{B_{j+1/2}^{n+1}}{\Delta v_{j+1/2}} - \frac{B_{j-1/2}^{n+1}}{\Delta v_{j-1/2}} \right) - C_j^{n+1} \right], \tag{2.2.26}$$

$$\gamma_j^{n+1} = \frac{\rho}{v_j^2} \frac{1}{\Delta v_j} \left(-\frac{A_{j-1/2}^{n+1}}{2} + \frac{B_{j-1/2}^{n+1}}{\Delta v_{j-1/2}} \right), \tag{2.2.27}$$

$$\delta_j^n = F_{j+1}^n \left[\frac{(1-\rho)}{v_j^2} \frac{1}{\Delta v_j} \left(\frac{A_{j+1/2}^n}{2} + \frac{B_{j+1/2}^n}{\Delta v_{j+1/2}} \right) \right]$$

$$F_j^n \left\{ \frac{1}{\Delta t} + \frac{(1-\rho)}{v_j^2} \left[\frac{1}{\Delta v_j} \left(\frac{A_{j+1/2}^n - A_{j-1/2}^n}{2} - \frac{B_{j+1/2}^n}{\Delta v_{j+1/2}} - \frac{B_{j-1/2}^n}{\Delta v_{j-1/2}} \right) - C_j^n \right] \right\} \tag{2.2.28}$$

$$+ F_{j-1}^n \left[\frac{(1-\rho)}{v_j^2} \frac{1}{\Delta v_j} \left(-\frac{A_{j-1/2}^n}{2} + \frac{B_{j-1/2}^n}{\Delta v_{j-1/2}} \right) \right]$$

$$+ \rho D_j^{n+1} + (1-\rho) D_j^n.$$

In order to linearize the system (2.2.24) we extrapolate the α, β, γ defined above to the new time step, $n + 1$. The method used to solve (2.2.24) is the standard tridiagonal method given by Richtmyer and Morton (1967).

The coefficient of the scattering loss term, Λ_a, can be obtained by solving the eigenvalue problem, (2.2.5), for a given θ_{LC}. For the eigenvalue of the equation corresponding to the first normal mode we can use the approximate value

$$\Lambda_a = (\log_{10} R_a)^{-1}, \tag{2.2.29}$$

where R_a is the effective mirror ratio for particles of type a and depends on the ambipolar potential $e\Phi$.

For electrons we have

$$R_e = R \left(1 - \frac{e\Phi}{\frac{1}{2} m_e v^2} \right)^{-1}, \tag{2.2.30}$$

where

$$R = \frac{B_{\max}}{B_0}, \qquad e\Phi = \tfrac{1}{2} m_e v_{cr}^2. \tag{2.2.31}$$

For ions we have

$$R_a = R \left(1 + \frac{Z_a e\Phi}{\frac{1}{2} m_a v^2} \right)^{-1}. \tag{2.2.32}$$

The potential $e\phi$ may either be specified or computed self-consistently. In the latter case one procedure for determining $e\phi$ is as follows. Let

$$Q^-(t) = n_e(t),\tag{2.2.33}$$

$$Q^+(t) = \sum_b Z_b n_b(t)\qquad (b \neq e)\tag{2.2.34}$$

(the sum taken over the ion species). At every time step, Q^- and Q^+ are computed and the difference $(Q^+ - Q^-)/Q^-$ is compared to a specified small number. If the difference exceeds this number, the v_{cr} is modified by an amount Δv_{cr} and the time step is repeated. This process is repeated until the condition is satisfied.

2.2.2. Applications

2.2.2.1. D–T and D–^{3}He mirror reactors

The lowest normal mode representation has been used in a parametric study of the figure of merit Q for mirror reactors employing D–T and D–^{3}He fuel cycles (Futch et al., 1972). In these investigations one considers a steady-state system in which the mirror-confined plasma is sustained by energetic neutral beam injection. The Q of the system is defined by

$$Q = \frac{\text{(thermonuclear power)}}{\text{(injection power required to maintain the plasma)}}$$

$$= \frac{\dfrac{1}{2}\sum_a \sum_b n_a n_b \langle \sigma v \rangle_{ab} E_{ab}}{\sum_a J_a E_a},\tag{2.2.35}$$

where n_a is the number density (particles cm^{-3}), J_a is the injected source current density (particles cm^{-3} s^{-1}), and E_a is the source energy for particles of species "a". The quantity $\langle \sigma v \rangle_{ab}$ is the product of the fusion cross section and relative speed, averaged over the distribution functions of the interacting particles, and E_{ab} is the energy released per fusion reaction. For given source parameters (J_a, E_a) and mirror ratio R_m, the veolocity distribution functions for all plasma species are obtained from (2.2.10) with source terms added and are used to calculate the thermonuclear reaction rate (Marx et al., 1976; Cordey et al., 1978)

$$n_a n_b \langle \sigma v \rangle_{ab} = \int d\vec{v}_a \int d\vec{v}_b f_a(\vec{v}_a) f_b(\vec{v}_b) \sigma_{ab}(v_{ab}) v_{ab},\tag{2.2.36}$$

where $v_{ab} = |\vec{v}_a - \vec{v}_b|$ is the relative speed. We neglect the anisotropic nature of the distribution functions in performing these Q calculations so the explicit form of the normal mode eigenfunction $M_a(\mu)$ is not needed.

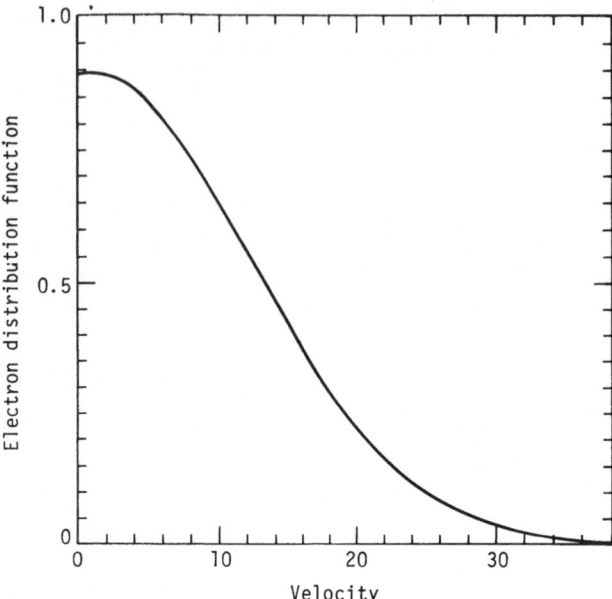

Figure 2.2. Electron
velocity distribution.

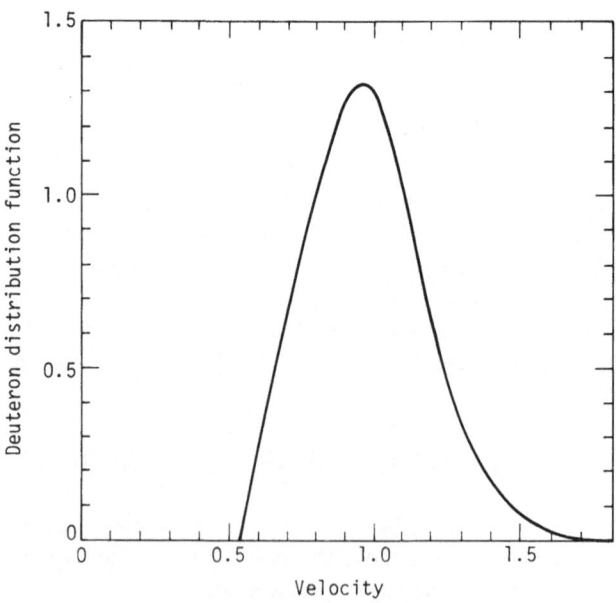

Figure 2.3. Deuteron
velocity distribution.

Figure 2.4. Triton veloc-
ity distribution.

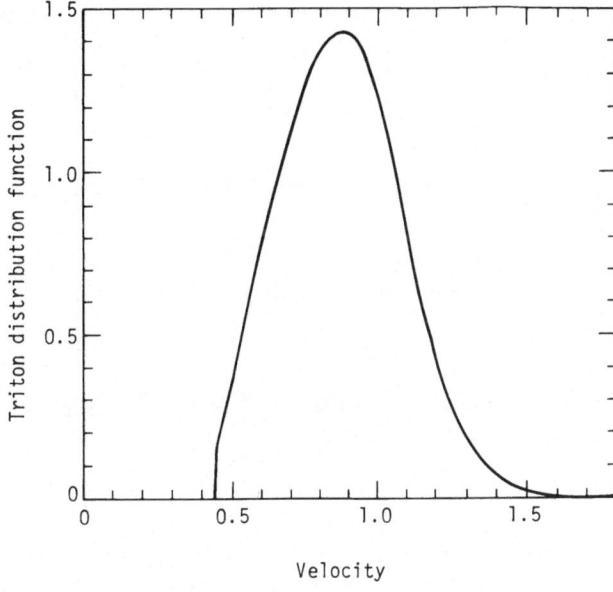

Figure 2.5. Lowest-
normal-mode angle
eigenfunction.

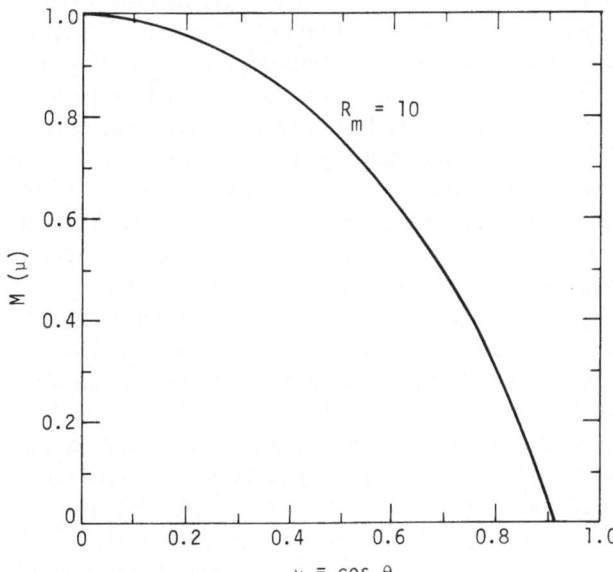

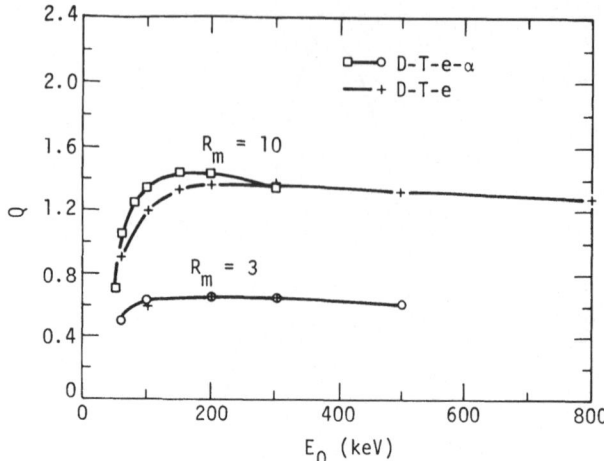

Figure 2.6. Mirror system Q versus injection energy ($E_{DT} = 22.4$ MeV).

To illustrate the nature of the solutions, we show some typical velocity distribution functions for electrons, deuterons, and tritons in Figs. 2.2–2.4. A typical normal mode angular distribution is shown in Fig. 2.5. The results of a parameter survey for the D–T mirror reactor fuel cycle are summarized in Fig. 2.6 where Q is given as a function of injection energy ($E_D = E_T = E_0$) and mirror ratio, with source currents adjusted so as to yield equal densities for the deuterons and tritons. Charged fusion reaction products (α-particles) in the four-species runs (D–T–E–α) are assumed to be adiabatically confined, thus producing additional heating as they slow down within the plasma. Their continuous creation is simulated by source terms similar in form to the ion source terms, but with current given by the D–T reaction rate, and with E_α equal to 3.5 MeV. The loss due to burn-up is included by subtracting the reaction rate from the D and T source terms. The results in Fig. 2.6 show that there is a very broad maximum in Q as a function of the D–T injection energy E_0, and α-particle confinement ceases to be beneficial beyond about $E_0 = 300$ keV. More details on these and other results can be found in Futch *et al.* (1972).

The lowest normal mode Fokker–Planck model has also been used in a study of two-component mirror reactors as described by Post *et al.* (1973). A modified loss term accounted for deuteron losses due to axial diffusion at low energies and mirror end losses at high energy. The bulk tritium and electron plasma was assumed to form a fixed Maxwellian background for the injected deuterons. Q values comparable to those for toroidal two-component systems (Dawson *et al.*, 1971) are obtained.

A computational study, using the lowest normal mode code, has been made (Mirin *et al.*, 1977) of the energy amplification factor Q to be expected in a toroidally linked mirror (TLM) reactor (Cordey and Watson, 1972, 1975).

Because of the uncertainty which exists over the effect of ambipolar electric fields on diffusional losses from TLM systems, two model loss terms have been considered. For each model, Q has been calculated for a range of values of the mirror ratio and the ion injection energy.

2.2.2.2. Two-component toroidal reactors

We illustrate the time-dependent or transient aspect of the solution to the kinetic equations for a multispecies plasma. We apply the model to the study of the energy multiplication factor in the two-component toroidal reactor (Dawson *et al.*, 1971; Berk *et al.*, 1975). In this application there are three plasma species (e–D–T), each of which is assumed to have an isotropic distribution in velocity space. We can also include the α-particles produced by D–T fusion reactions. Energetic deuterons are injected into an ohmic-heated tritium plasma in a tokamak. We follow the evolution of the initially peaked deuteron distribution and calculate the energy produced via D–T fusion reactions as the deuterons slow down in the background plasma. In this instance there are no loss terms in the kinetic equations (i.e., no loss cone in velocity space) so we effectively assume that the particle and energy confinement times are long compared to the slowing-down time for the deuterons. The figure of merit for the system is the energy multiplication factor, F, defined by

$$F = \frac{\text{(fusion energy produced)}}{\text{(initial energy in the deuterons)}} = \frac{\int_0^t dt\, n_D n_T \langle \sigma v \rangle (t) E_F}{n_D \bar{E}_D(0)}, \qquad (2.2.37)$$

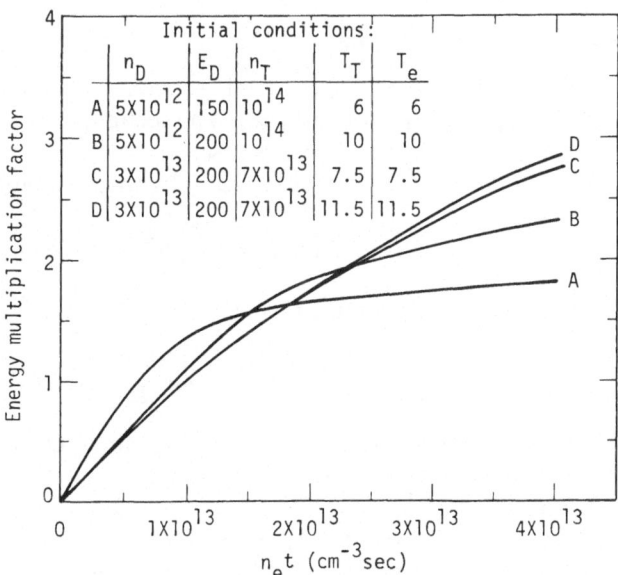

Figure 2.7. TCT energy multiplication (curve D includes alphas).

where $\bar{E}_D(0)$ is the mean energy per deuteron at $t = 0$ and $\langle \sigma v \rangle$ is the D–T fusion rate parameter which depends on the detailed shape of the distributions.

Results for the energy multiplication factor F in several illustrative cases are given in Fig. 2.7. For times long compared to the slowing-down time of the deuterons, the system approaches thermal equilibrium so that $\langle \sigma v \rangle$ becomes constant in time and F increases linearly. This is clearly seen in Fig. 2.7. To obtain an F-value which does *not* depend on the time interval chosen in (2.2.37), one can define a "transient" contribution to F by subtracting the asymptotic steady-state component of the fusion power from the integrand in (2.2.37). This procedure yields well-defined F-values which are more nearly comparable to those obtained by Dawson *et al.* (1971).

2.3. Solution in a Two-Dimensional Velocity Space Using an Expansion in Pitch-Angle

The separated solution approach described in Section 2.2 can be generalized in two ways. First, the assumption that the Rosenbluth potentials g_a and h_a are isotropic is not made, in which case the expansion in Legendre polynomials given by (2.1.10) and (2.1.11) is used. Second, the angular dependence of the solution is represented by a sum of normal modes rather than just the lowest normal mode.

2.3.1. Numerical methods

It will be assumed throughout this section that our confinement system is a mirror trap with mirror ratio R_m. The technique of eigenfunction expansion, however, is not restricted to this case.

We represent the solutions of (2.1.7) by the orthogonal series

$$f_a(v, \mu, t) = \sum_{l=1}^{\infty} U_l^a(v, t) M_l^a(\mu),\tag{2.3.1}$$

where $M_l^a(\mu)$ are eigenfunctions of

$$(1 - \mu^2)\frac{d^2 M_l^a}{d\mu^2} - 2\mu\frac{dM_l^a}{d\mu} + \lambda_l^a M_l^a = 0,\tag{2.3.2}$$

which is Legendre's equation on the domain $-\cos\theta_{LC} \leq \mu \leq \cos\theta_{LC}$, where θ_{LC} is the loss cone angle defined by (2.1.55).

In this treatment we do not assume that g_a and h_a are isotropic, so (2.1.7) is not separable. However, we do use the fact that the functions $M_l^a(\mu)$ form a complete orthonormal set on the domain $-\mu_{LC} \leq \mu \leq \mu_{LC} = \cos\theta_{LC}$; i.e., for

$k \neq l$

$$\int_{-\mu_{LC}}^{\mu_{LC}} M_l^a(\mu) M_k^a(\mu)\, d\mu = 0; \qquad \int_{-\mu_{LC}}^{\mu_{LC}} [M_l^a(\mu)]^2\, d\mu = 1. \tag{2.3.3}$$

For the purposes of computation, we consider a finite number, N, of normal modes, $M_k^a(\mu)$, for each species. We obtain these functions by the numerical solution of (2.3.2), as a two-point eigenvalue problem, with $M_k^a(\mu) = 0$ at the endpoints, for $k = 1, 2, \ldots, N$.

If we substitute the right-hand side of (2.3.1) into (2.1.7), multiply by $M_k^a(\mu)$ and integrate with respect to μ, making use of (2.3.3), we obtain the following equation for $U_k^a(v, t)$

$$\frac{\partial U_k^a}{\partial t} = \sum_{l=1}^{N} \left\{ \frac{1}{v^2} \frac{\partial}{\partial v} \left[\alpha_{kl}^a U_l^a + \beta_{kl}^a \frac{\partial U_l^a}{\partial v} \right] - \frac{\gamma_{kl}^a}{v^2} U_l^a \right\} \tag{2.3.4}$$

for $k = 1, \ldots, N$, where

$$\alpha_{kl}^a = \Gamma_a \int_{-\mu_{LC}}^{\mu_{LC}} d\mu\, M_k^a(\mu) \left\{ M_l^a(\mu) \left[-v^2 \frac{\partial h_a}{\partial v} + \frac{v^2}{2} \frac{\partial^3 g_a}{\partial v^3} + v \frac{\partial^2 g_a}{\partial v^2} \right. \right.$$
$$\left. - \frac{1}{2v}(1 - \mu^2) \frac{\partial^2 g_a}{\partial \mu^2} - \frac{\partial g_a}{\partial v} + \frac{\mu}{v} \frac{\partial g_a}{\partial \mu} \right] \tag{2.3.5}$$
$$\left. + \frac{\partial}{\partial \mu} \left[(1 - \mu^2) M_l^a \left(\frac{\partial^2 g_a}{\partial \mu\, \partial v} - \frac{1}{v} \frac{\partial g_a}{\partial \mu} \right) \right] \right\},$$

$$\beta_{kl}^a = \Gamma_a \int_{-\mu_{LC}}^{\mu_{LC}} d\mu \left[M_k^a(\mu) M_l^a(\mu) \frac{v^2}{2} \frac{\partial^2 g_a}{\partial v^2} \right], \tag{2.3.6}$$

$$\gamma_{kl}^a = -\Gamma_a \int_{-\mu_{LC}}^{\mu_{LC}} d\mu\, M_k^a(\mu) \left\{ \frac{\partial}{\partial \mu} \left[M_l^a(\mu) \left(-(1 - \mu^2) \frac{\partial h_a}{\partial \mu} \right. \right. \right.$$
$$\left. + \frac{\mu}{2v^2}(1 - \mu^2) \frac{\partial^2 g_a}{\partial \mu^2} + \frac{\mu}{v} \frac{\partial g_a}{\partial v} + \frac{(1 - \mu^2)}{v} \frac{\partial^2 g_a}{\partial \mu\, \partial v} - \frac{1}{v^2} \frac{\partial g_a}{\partial \mu} \right) \right]$$
$$\left. + \frac{\partial^2}{\partial \mu^2} \left[M_l^a(\mu) \left(\frac{(1 - \mu^2)^2}{2v^2} \frac{\partial^2 g_a}{\partial \mu^2} + \frac{(1 - \mu^2)}{2v} \frac{\partial g_a}{\partial v} - \frac{\mu(1 - \mu^2)}{2v^2} \frac{\partial g_a}{\partial \mu} \right) \right] \right\}. \tag{2.3.7}$$

We simplify (2.3.7) by making use of (2.3.2); hence

$$\gamma_{kl}^a = -\Gamma_a \int_{-\mu_{LC}}^{\mu_{LC}} d\mu\, M_k^a(\mu) \left\{ \frac{\partial M_l^a}{\partial \mu} \left[\frac{2(1 - \mu^2)}{v} \frac{\partial^2 g_a}{\partial \mu\, \partial v} - \frac{7\mu(1 - \mu^2)}{2v^2} \frac{\partial^2 g_a}{\partial \mu^2} \right. \right.$$
$$\left. + \frac{(1 - \mu^2)^2}{v^2} \frac{\partial^3 g_a}{\partial \mu^3} + \frac{2(\mu^2 - 1)}{v^2} \frac{\partial g_a}{\partial \mu} - (1 - \mu^2) \frac{\partial h_a}{\partial \mu} \right]$$
$$+ M_l^a \left[-\frac{3\mu}{v} \frac{\partial^2 g_a}{\partial \mu\, \partial v} + \left(\frac{15\mu^2 - 7}{2v^2} - \frac{\lambda_l^a(1 - \mu^2)}{2v^2} \right) \frac{\partial^2 g_a}{\partial \mu^2} \right.$$

$$
-\frac{4\mu(1-\mu^2)}{v^2}\frac{\partial^3 g_a}{\partial\mu^3} + \frac{3(1-\mu^2)}{2v}\frac{\partial^3 g_a}{\partial\mu^2\,\partial v} + \frac{(1-\mu^2)^2}{2v^2}\frac{\partial^4 g_a}{\partial\mu^4}
$$

$$
+\left(\frac{3\mu}{v^2} + \lambda_l^a\frac{\mu}{2v^2}\right)\frac{\partial g_a}{\partial\mu} - \frac{\lambda_l^a}{2v}\frac{\partial g_a}{\partial v} + 2\mu\frac{\partial h_a}{\partial\mu} - (1-\mu^2)\frac{\partial^2 h_a}{\partial\mu^2}\bigg]\bigg\}. \tag{2.3.8}
$$

The functions g_a and h_a in (2.3.5)–(2.3.8) are represented by the expansions (2.1.10) and (2.1.11), where (2.1.12) and (2.1.13) make use of

$$
V_j^a(v,t) = \frac{\displaystyle\sum_{l=1}^{N} U_l^a(v,t)\int_{-1}^{+1} M_l^a(\mu)P_j(\mu)\,d\mu}{\displaystyle\int_{-1}^{+1}[P_j(\mu)]^2\,d\mu}. \tag{2.3.9}
$$

All of the derivatives of g_a and h_a with respect to v and μ can be carried out analytically, as described by (2.1.25)–(2.1.34). The coefficients (2.3.5)–(2.3.8) of (2.3.4) can be expressed analytically in terms of the four time-dependent functionals defined by (2.1.25)–(2.1.28), i.e.,

$$
M_j(V_j^a) = \int_v^\infty V_j^a(v',t)(v')^{1-j}\,dv', \tag{2.3.10}
$$

$$
N_j(V_j^a) = \int_0^v V_j^a(v',t)(v')^{2+j}\,dv', \tag{2.3.11}
$$

$$
E_j(V_j^a) = \int_0^v V_j^a(v',t)(v')^{4+j}\,dv', \tag{2.3.12}
$$

$$
R_j(V_j^a) = \int_v^\infty V_j^a(v',t)(v')^{3-j}\,dv', \tag{2.3.13}
$$

and five definite integrals

$$
(S_1)_{jkl}^{ab} = \int_{-\mu_{LC}}^{+\mu_{LC}} \frac{dP_j}{d\mu}M_k^a(\mu)\frac{dM_l^b}{d\mu}(1-\mu^2)\,d\mu, \tag{2.3.14}
$$

$$
(S_2)_{jkl}^{ab} = \int_{-\mu_{LC}}^{+\mu_{LC}} P_j(\mu)M_k^a(\mu)M_l^b(\mu)\,d\mu, \tag{2.3.15}
$$

$$
(S_3)_{jkl}^{ab} = \int_{-\mu_{LC}}^{+\mu_{LC}} \frac{dP_j}{d\mu}M_k^a(\mu)\frac{dM_l^b}{d\mu}\,d\mu, \tag{2.3.16}
$$

$$
(S_4)_{jkl}^{ab} = \int_{-\mu_{LC}}^{+\mu_{LC}} P_j(\mu)M_k^a(\mu)\frac{dM_l^b}{d\mu}\mu\,d\mu, \tag{2.3.17}
$$

$$
(S_5)_{jkl}^{ab} = \int_{-\mu_{LC}}^{+\mu_{LC}} \frac{dP_j}{d\mu}M_k^a(\mu)M_l^b(\mu)\mu\,d\mu. \tag{2.3.18}
$$

The integrals given by (2.3.14)–(2.3.18) are computed using (2.1.55) for the loss cone angle, which does not depend on the ambipolar potential and is time-independent and species-independent; hence, the integrals are time-independent.

The procedure described so far in Section 2.3.1 is rigorous and consistent, and no approximations have been made, other than using a finite number of normal modes and Legendre polynomials in the representation of the solution and the Rosenbluth potentials. However, the effect of a confining potential, which is introduced in Section 2.1.3, has not yet been included. In order to consider the ambipolar potential we should use (2.1.56), instead of (2.1.55), for the loss cone angle, θ_{LC}; however this expression is time-dependent and species-dependent. Using (2.1.56) would greatly complicate the formalism described as the normal modes, $M_l^a(\mu)$, would no longer be time-independent and the solution procedure given by (2.3.1)–(2.3.4) would not be valid.

In order to keep the present formalism we use the time-independent form (2.1.55) to compute the normal modes defined by (2.3.2). This also maintains the time-independence of the integrals (2.3.14)–(2.3.18). However, we include the effect of the ambipolar potential in the loss term of (2.3.4) by modifying the eigenvalues, λ_l^a. The eigenvalues λ_l^a which appear in (2.3.8) do depend on the "effective" mirror ratio, R_a, given in (2.1.56b). Note that this formula applies only if "a" is an ion and $v^2 \geq v_{ca}^2 = Z_a|e\phi|/\frac{1}{2}m_a(R_m - 1)$ or if "a" is an electron and $v^2 \geq v_{ce}^2 = |e\phi|/\frac{1}{2}m_e$. We extend the definition of R_a to all velocities as $R_a = 1$ if "a" is an ion and $v^2 \leq v_{ca}^2$, and $R_e = \infty$ if $v^2 \geq v_{ce}^2$. The smallest eigenvalue λ_1^a is given by

$$\lambda_1^a \approx 1/\log_{10} R_a, \tag{2.3.19}$$

and the larger eigenvalues λ_l^a ($l \neq 1$) are obtained through piecewise linear approximations of λ_l^a/λ_1^a as functions of $\log_{10} R_a$. The distribution function U_l^a is set to 0 at those points where λ_l^a is infinite. We also ignore all anisotropic components of the Rosenbluth potentials for all non-ion–ion interactions; i.e., we set $(S_{1-5})_{jkl}^{ab} = 0$ if $j > 0$ and "a" or "b" equals "e".

Equations (2.3.4) are solved on a finite-difference mesh $\{v_j\}_{j=1}^J$, where $v_1 = 0$ and $v_2 \leq v_J/(J - 1)$. The value of v_2 determines a unique mesh ratio $r \geq 1$ satisfying

$$v_{j+1} - v_j = r(v_j - v_{j-1}), \qquad j = 2, \ldots, J - 1. \tag{2.3.20}$$

The purpose of incorporating a nonuniform mesh of this type is to assure adequate representation of the low-velocity ions and the high-velocity electrons.

An implicit algorithm is used to integrate the vector system (2.3.4). The spatial and temporal difference approximations are analogous to those of Section 2.2.1 for (2.2.17). For each species "a" we obtain a tridiagonal system of the form

$$-A_j U_{j+1}^{n+1} + B_j U_j^{n+1} - C_j U_{j-1}^{n+1} = D_j, \tag{2.3.21}$$

where A_j, B_j, and C_j are nondiagonal $N \times N$ matrices, D_j is an N-vector, and

U_j^n is an N-vector which represents $U_k^a(v_j, n\Delta t)$, $k = 1, \ldots, N$. The method of solution is described in Richtmyer and Morton (1967).

The procedure for determining the ambipolar potential is the same as that described in Section 2.2.1. The lowest eigenvalues, λ_1^a, are given by (2.2.29)–(2.2.32), and the larger eigenvalues of (2.3.2) are modified as described earlier (after (2.3.19)).

2.3.2. Applications

We consider a mirror-confined deuterium plasma in which the injection energy of the deuterons is 100 keV. The ambipolar potential is computed self-consistently as described earlier. We calculate the confinement figure of merit

$$n\tau = n_i^2/J_i, \qquad (2.3.22)$$

where n_i is the ion number density (particles cm^{-3}) and J_i is the injected ion source current density (particles cm^{-3} s^{-1}). We calculate $n\tau$ as a function of the angular shape of the injected beam, the mirror ratio, R, the number of even Legendre polynomials, M, in the Rosenbluth potential expansions, and the

Table 2.1. Comparisons of confinement parameter values.

Angular source	R	M	N	Code	$n\tau$ ($\times 10^{13}$)
Normal mode	3	1	1	expansion	1.03
Normal mode	3	2	1	expansion	1.13
Normal mode	3	3	1	expansion	1.14
Normal mode	3	4	1	expansion	1.14
Normal mode	3	5	1	expansion	1.14
Normal mode	3	2	4	expansion	1.09
Normal mode	3	5	—	difference	1.02
Narrow ($e^{-10\mu^2}$)	3	2	4	expansion	1.24
Narrow	3	5	—	difference	1.22
Normal mode	10	1	1	expansion	2.07
Normal mode	10	2	1	expansion	2.34
Normal mode	10	3	1	expansion	2.34
Normal mode	10	4	1	expansion	2.34
Normal mode	10	5	1	expansion	2.34
Normal mode	10	2	4	expansion	2.29
Normal mode	10	5	—	difference	2.24
Narrow ($e^{-10\mu^2}$)	10	1	2	expansion	2.48
Narrow	10	1	3	expansion	2.34
Narrow	10	1	4	expansion	2.36
Narrow	10	2	2	expansion	2.86
Narrow	10	2	3	expansion	2.58
Narrow	10	2	4	expansion	2.61
Narrow	10	3	2	expansion	2.86
Narrow	10	5	—	difference	2.65

number of normal modes, N, used in (2.3.1) to represent the solution of (2.1.7). For the angular shape function of the source, two models are used: the lowest normal model solution, $M_1^a(\mu)$, of 2.3.2, corresponding to the mirror ratio R; and a narrow source, $e^{-10\mu^2}$, which is appropriate for perpendicular injection into the mirror device.

Results of this study are shown in Table 2.1, where we also include results from the two-dimensional finite-difference code discussed in Section 2.4.

From Table 2.1 we see that, for problems in which the source is proportional to the lowest normal mode, reasonably accurate results may be obtained by considering only one normal mode (i.e., $N = 1$) and by considering only the terms in (2.1.10) and (2.1.11) proportional to $P_0(\mu)$ and $P_2(\mu)$. (The term corresponding to $P_1(\mu)$ vanishes due to symmetry about $\mu = 0$.) For narrow source problems there is good agreement between the two-dimensional difference code of Section 2.4 and the expansion code with $M = 2$ and $N = 4$.

2.3.3. Modification of the lowest normal mode code of Section 2.2 to include anisotropic Rosenbluth potentials

The results of Table 2.1 show that, for normal mode sources, the expansion code gives better results for $N = 1$ with $M \geq 2$, and that the results change little with $M > 2$. It was decided to revise the one-dimensional, lowest normal mode code of Section 2.2 by including the $P_2(\mu)$ terms in the Rosenbluth potentials, using the formalism of Section 2.3.1 given by (2.3.4)–(2.3.18), but with $N = 1$, $M = 2$. This is now the standard version of that code (Mirin, 1974).

Some typical cases of the parametric study of the figure of merit Q for mirror reactors employing the D–T and D–^{3}He fuel cycles (Futch et al., 1972) were repeated using the modified one-dimensional code, the expansion code, and

Table 2.2. Comparisons of Q values.
$$(R_m = 10, E_0 = 100 \text{ keV})$$

Code description	Plasma species			
	D–e–T	D–e–T–α	D–e–^{3}He	D–e–^{3}He–α–p
one-dimensional code; P_0 only	1.22	1.39	0.244	0.264
one-dimensional code; P_0 and P_2	1.44	1.68	0.289	0.312
two-dimensional difference code; normal mode source	1.38	1.61	0.278	0.301
two-dimensional difference code; narrow source	1.71	1.99	0.337	0.365

the two-dimensional difference code described in Section 2.4. The results in Fig. 2.6 and Futch *et al.* (1972) were calculated using only the $P_0(\mu)$ term. The effect of including the next nonzero term in (2.1.10) and (2.1.11), the $P_2(\mu)$ term, is shown in Table 2.2. One sees that Q increases by about 20%. Additional terms in (2.1.10) and (2.1.11), using the expansion code, yield smaller corrections. Results from the two-dimensional difference code are also included in Table 2.2 and good agreement is obtained when the $P_2(\mu)$ corrections are included in the one-dimensional model.

The Q calculations for a toroidally linked mirror (TLM) reactor, which were mentioned in Section 2.2.2 and reported in Mirin *et al.* (1977), were performed with the modified one-dimensional code which includes the $P_2(\mu)$ correction.

2.4. Solution Using Finite-Differences in a Two-Dimensional Velocity Space

In this section we describe our standard two-dimensional, multispecies, non-linear Fokker–Planck code which includes the Fokker–Planck solver and much additional physics as described in Sections 2.1 and 2.4.2. The Fokker–Planck package (FPPAC) has been optimized for the CRAY-1 vector computer (McCoy *et al.*, 1979) and is available through the CPC program library (McCoy *et al.*, 1981).

An earlier two-dimensional finite-difference code was developed (Killeen and Marx, 1970) which solved the unseparated Fokker–Planck equation in v and θ for a single ion species, under the assumption that the electrons can be represented by a Maxwellian distribution function with loss cone removed. The ion equation was solved using the alternating-direction implicit (ADI) finite-difference scheme on a variable mesh in v and θ, and the coefficients, containing moments over both ion and electron distribution functions, recomputed at each ion timestep. The number density and mean energy, characterizing the Maxwellian distribution of electrons, were computed on a faster time scale by solving a pair of ordinary differential equations that included the effects of ion–electron interactions and electron endlosses and injection conditions. The ambipolar potential was determined by iteration on the electron equations in such a way as to equalize the positive and negative charge densities. The Rosenbluth potentials were computed numerically in their integral form, (2.1.3) and (2.1.4), with the singular integral, h_a, transformed because of the single species nature of the model. For a multispecies code it is much more convenient to use the expansions given by (2.1.10) and (2.1.11), and all the derivatives can be performed analytically as described in (2.1.31)–(2.1.34).

The reasons that a two-dimensional finite-difference code was developed, when the two-dimensional expansion code of Section 2.3 was already available, are speed and consistency. For a narrow angular source, several angular normal modes are needed to represent the distribution function, and when comparisons were made with the single-species, two-dimensional finite-difference code (Killeen and Marx, 1970), the finite-difference code in v and θ was much faster. Furthermore, as discussed in Section 2.3.1, for mirror problems with an ambipolar potential, there is a lack of consistency, as the normal modes are computed neglecting the ambipolar potential. The code to be described in Section 2.4.1 resolves this problem and is also much faster than the expansion code when several normal modes are used. For these reasons it has become our standard Fokker–Planck code for plasmas in uniform magnetic fields.

2.4.1. Numerical methods

2.4.1.1. Basic equations

In this section we use the conservative form of the Fokker–Planck equation given by (2.1.14)–(2.1.16) for each species, i.e.,

$$\frac{1}{\Gamma_a}\left(\frac{\partial f_a}{\partial t}\right)_c = \frac{1}{v^2}\frac{\partial G_a}{\partial v} + \frac{1}{v^2\sin\theta}\frac{\partial H_a}{\partial\theta}, \tag{2.4.1}$$

where

$$G_a = A_a f_a + B_a\frac{\partial f_a}{\partial v} + C_a\frac{\partial f_a}{\partial\theta}, \tag{2.4.2}$$

$$H_a = D_a f_a + E_a\frac{\partial f_a}{\partial v} + F_a\frac{\partial f_a}{\partial\theta}. \tag{2.4.3}$$

The coefficients A_a, B_a, C_a, D_a, E_a, and F_a are given by (2.1.17)–(2.1.22). The form of the Rosenbluth potentials, $g_a(v, \theta, t)$ and $h_a(v, \theta, t)$, which is used is given by (2.1.10) and (2.1.11) with $\mu = \cos\theta$. Using the functionals defined by (2.1.25)–(2.1.28), the coefficients of g_a and h_a, defined by (2.1.12) and (2.1.13), are expressed as (2.1.29) and (2.1.30). The derivatives of g_a and h_a, which are needed in calculating the coefficients (2.1.17)–(2.1.22), are performed analytically making use of (2.1.31)–(2.1.34) and the differentiation formulas for Legendre polynomials.

2.4.1.2. Boundary conditions

The boundary conditions for (2.4.1) are described in Section 2.1.3.

2.4.1.3. Mesh definition

The independent variables v and θ are represented by meshes $\{v_j\}_{j=1}^J$ and $\{\theta_i\}_{i=1}^I$, respectively. These meshes do not have to be uniform. The first velocity meshpoint, v_1, will always equal 0. The θ-domain may be of two types:

(1) Assumed symmetry of the distribution functions about $\theta = \pi/2$, in which case $\theta_I = \pi/2$. This case utilizes an extra reflection point $\theta_{I+1} = \pi - \theta_{I-1}$.

(2) No assumed symmetry of the distribution functions. Here, the θ-mesh is taken to be symmetric about $\theta = \pi/2$, with a meshpoint right at $\theta = \pi/2$, so that I must be odd. In either case, θ_1 need not be equal to 0.

2.4.1.4 *Numerical integration in velocity space*

Numerical integration in velocity space is accomplished through a variation of trapezoidal integration. In the v direction,

$$\int f(v)v^2 \, dv \approx \sum_{j=1}^{J} a_j f(v_j), \tag{2.4.4}$$

where

$$a_1 = \frac{v_2^3}{24}, \tag{2.4.5}$$

$$a_j = v_j^2 \left(\frac{v_{j+1} - v_{j-1}}{2} \right), \qquad 1 < j < J, \tag{2.4.6}$$

$$a_J = v_J^2 \left(\frac{v_J - v_{J-1}}{2} \right). \tag{2.4.7}$$

Equation (2.4.5) assumes $f(v)$ is constant between $v = 0$ and $v = v_2/2$, so that

$$\int_{v_1=0}^{v_2/2} f(v)v^2 \, dv = f(v_1)\left(\frac{v_2}{2} \right)^3 \Big/ 3. \tag{2.4.8}$$

This procedure is more accurate than the straightfoward scheme having $a_1 = v_1^2 \left(\frac{v_2 - v_1}{2} \right) = 0$. In the θ-direction,

$$\int f(\theta) \sin \theta \, d\theta \approx \sum_{i=1}^{I} b_i f(\theta_i), \tag{2.4.9}$$

where

$$b_1 = \begin{cases} \sin \theta_1 \left(\dfrac{\theta_2 - \theta_1}{2} \right) & \text{if } \theta_1 \neq 0, \\[2ex] \theta_2^2/8 & \text{if } \theta_1 = 0, \end{cases} \tag{2.4.10}$$

$$b_i = \sin \theta_i \left(\frac{\theta_{i+1} - \theta_{i-1}}{2} \right), \qquad 1 < i < I, \tag{2.4.11}$$

$$b_I = \begin{cases} \sin \theta_I \left(\dfrac{\theta_I - \theta_{I-1}}{2} \right) & \text{if } \theta_I \neq \pi, \\[2ex] (\pi - \theta_{I-1})^2/8 & \text{if } \theta_I = \pi, \end{cases} \tag{2.4.12}$$

Thus, the density of species "a" is computed as

$$n_a = 2\pi \sum_{i,j} a_j b_i f_a(v_j, \theta_i), \qquad (2.4.13)$$

and the energy density of species "a" is

$$n_a E_a = \pi m_a \sum_{i,j} a_j b_i v_j^2 f_a(v_j, \theta_i). \qquad (2.4.14)$$

2.4.1.5. Computation of Fokker–Planck coefficients

The collision operator, $(\partial f_a/\partial t)_c$, may be written as a sum

$$\left(\frac{\partial f_a}{\partial t}\right)_c = \sum_b \left(\frac{\partial f_a}{\partial t}\right)_c^b, \qquad (2.4.15)$$

where $(\partial f_a/\partial t)_c^b$ represents collisions of species "a" with species "b". In computing $(\partial f_a/\partial t)_c^b$ for a general species "b", the procedure is that described by (2.1.8)–(2.1.34), with the proviso that the sum over "b" in (2.1.10) and (2.1.11) be replaced by the single index "b". The only numerical approximations come about in the integrals of (2.1.9) and (2.1.25)–(2.1.28). Equation (2.1.9) is approximated using (2.4.9)–(2.4.12). Equations (2.1.25)–(2.1.28) are approximated by

$$\int_{v_k}^{v_l} f(v)\, dv = \sum_{j=k}^{l-1} \left(\frac{v_{j+1} - v_j}{2}\right)(f(v_j) + f(v_{j+1})). \qquad (2.4.16)$$

In order to take full advantage of the CRAY vector capabilities, it is necessary to organize the calculation of the coefficients in an optimal way. An explanation of the do-loop structure for the CRAY version of the code follows.

Since the structure of each of the six coefficients is quite similar, the detailed evaluation of only one of them, namely C_a (see (2.1.19)), is considered.

Let

$$\tilde{C}_l^b(v_j, \theta_i) = -\frac{1}{2v_j} B_l^b(v_j)\frac{\partial}{\partial \theta} P_l(\cos \theta_i)$$
$$+ \frac{1}{2}\frac{\partial}{\partial v} B_l^b(v_j)\frac{\partial}{\partial \theta} P_l(\cos \theta_i), \qquad (2.4.17)$$

C_a may be expressed (using (2.1.19) and (2.1.10)) in terms of the various $\tilde{C}_l^b$ coefficients as

$$C_a(v_j, \theta_i) = \sum_b \ln \Lambda_{ab} \left(\frac{Z_b}{Z_a}\right)^2 \sum_{l=0}^{M} \tilde{C}_l^b(v_j, \theta_i). \qquad (2.4.18)$$

Note that the Legendre series is truncated at $M + 1$ polynomials. Coefficients $\tilde{A}_l^b$ through $\tilde{F}_l^b$ may be similarly defined and expressed.

The expressions for the coefficients $\tilde{A}_l^b$ through $\tilde{F}_l^b$ are complicated (e.g., (2.4.17)) and these coefficients must be evaluated at all of the meshpoints. In fact, the computation of these coefficients represents a high percentage of the total number of arithmetic operations required; consequently, it is necessary to calculate these coefficients as rapidly as possible. This is accomplished through a series of nested "do-loops" which the CRAY FORTRAN (CFT) compiler vectorizes.

The coefficients $\tilde{A}_l^b(v_j, \theta_i)$ through $\tilde{F}_l^b(v_j, \theta_i)$ are functions of the four indices, "b", "l", "j", and "i", and efficiency considerations dictate that the species index "b" and the Legendre index "l" form the outermost "do-loops", since "b" and "l" are never large. Within these two loops one must evaluate (2.4.17) for all (v_j, θ_i) meshpoints. The procedure may be outlined as follows:

(1) Calculate the Legendre coefficients $V_l^b(v_j, t)$ for $j = 1, \ldots, J$ ((2.1.9)). This involves the calculation of J integrals over θ, and may be calculated in two nested loops with the outermost loop over θ ("i"). This will permit compiler vectorization of the entire procedure, since the inner loop over v ("j") is not recursive.

(2) Evaluate the functionals $M_l(V_l^b)(v_j)$, $N_l(V_l^b)(v_j)$, $R_l(V_l^b)(v_j)$, $E_l(V_l^b)(v_j)$ for $j = 1, \ldots, J$ (see (2.1.25)–(2.1.28)). This is accomplished in two separate loops over "j". For the functional $N_l(V_l^b)(v_j)$, the first loop calculates the temporary array $\text{TEM}(j) = V_l^b(v_j)v_j^{(2+l)}\Delta v_j$, where Δv_j is a mesh increment, while a second separate loop adds these temporaries together to form the functional $N_l(V_l^b)(v_j)$. This second loop does not vectorize since it is recursive.

(3) Determine the coefficients $B_l^b(v_j)$, $(\partial/\partial v)B_l^b(v_j)$, etc. for $j = 1, \ldots, J$ ((2.1.29)–(2.1.34)). This is accomplished through a single vectorizable loop over "j".

(4) Determine $\tilde{A}_l^b(v_j, \theta_i)$ through $\tilde{F}_l^b(v_j, \theta_i)$ for $i = 1, \ldots, I; j = 1, \ldots, J$. It is evident from (2.4.17) that the procedure is vectorizable. The theta ("i") loop is chosen on the outside, since it is normally much smaller than the velocity ("j") loop. The Legendre polynomials and their derivatives are time-independent and are stored quantities. As the coefficients $\tilde{A}_l^b - \tilde{F}_l^b$ are calculated, they are simultaneously summed over "l" since it is the sum that is required in (2.4.18).

(5) Evaluate the contribution of the bth species to the Fokker–Planck coefficients $A_a(v_j, \theta_i, t)$ through $F_a(v_j, \theta_i, t)$ for all species "a" (see (2.4.18)). This is accomplished within the two outer loops over "b" and "l" and within a third loop over "a". The procedure easily vectorizes.

Up to this point the assumption has been made that "b" is a "general" species represented by a two-dimensional velocity space distribution function. Recalling that FPPAC allows fixed background Maxwellian species in addition to general species, the procedure when "b" is a Maxwellian simplifies somewhat. Letting n_b denote its density and T_b its temperature,

$$\left(\frac{\partial f_a}{\partial t}\right)_c^b = \frac{\Gamma_a}{v^2}\left(\frac{\partial}{\partial v}\left(A_a^b f_a + B_a^b \frac{\partial f_a}{\partial v}\right) + \frac{1}{\sin\theta}\frac{\partial}{\partial \theta}\left(F_a^b \frac{\partial f_a}{\partial \theta}\right)\right), \qquad (2.4.19)$$

with

$$A_a^b = \left(\frac{Z_b}{Z_a}\right)^2 \ln \Lambda_{ab} \frac{m_a}{m_b} n_b \left(-\frac{v}{v_b} \sqrt{\frac{2}{\pi}} e^{-v^2/2v_b^2} + \phi\left(\frac{v}{\sqrt{2}v_b}\right)\right), \tag{2.4.20}$$

$$B_a^b = \frac{v_b^2}{v} \frac{m_b}{m_a} A_a^b, \tag{2.4.21}$$

$$F_a^b = \left(\frac{Z_b}{Z_a}\right)^2 \ln \Lambda_{ab} \frac{n_b \sin \theta}{2v} \left(\frac{v_b}{v} \sqrt{\frac{2}{\pi}} e^{-v^2/2v_b^2} + \left(1 - \frac{v_b^2}{v^2}\right)\phi\left(\frac{v}{\sqrt{2}v_b}\right)\right). \tag{2.4.22}$$

Here, n_b is the density of species "b", $v_b^2 = T_b/m_b$, and ϕ is the error function

$$\phi(x) = 2/\sqrt{\pi} \int_0^x e^{-y^2} \, dy. \tag{2.4.23}$$

For $x < 0.6$, $\phi(x)$ is approximated by a power series

$$\phi(x) \approx \sum_{k=0}^{6} (-1)^k \frac{x^{2k+2}}{k!(2k+1)}, \tag{2.4.24}$$

and for $x \geq 0.6$

$$\phi(x) \approx 1 - \frac{2}{\sqrt{\pi}} e^{-x^2} \sum_{k=1}^{5} A_k \eta^k, \tag{2.4.25}$$

with

$$\eta = 1/(1 + px), \tag{2.4.26}$$

and $p = 0.3275911$, $A_1 = 0.225836846$, $A_2 = -0.252128668$, $A_3 = 1.25969513$, $A_4 = -1.287822453$ and $A_5 = 0.94064607$ (Hastings, 1955).

2.4.1.6. Augmented equations

There are many situations in which it is desirable to model not only Coulomb collisions but additional physics as well (e.g., RF heating, electric field acceleration). This is provided for to the extent that the equations to be time-advanced are of the general form (dropping species identifier "a")

$$\frac{\partial f}{\partial t} = \frac{1}{v^2} \frac{\partial}{\partial v}\left(Af + B\frac{\partial f}{\partial v} + C\frac{\partial f}{\partial \theta}\right)$$
$$+ \frac{1}{v^2 \sin \theta} \frac{\partial}{\partial \theta}\left(Df + E\frac{\partial f}{\partial v} + F\frac{\partial f}{\partial \theta}\right) + Kf + J, \tag{2.4.27}$$

where A, B, C, D, E, F, J, and K are arbitrary functions of v and θ. The coefficients A, B, C, D, E, F include contributions from the collision operator, $(\partial f/\partial t)_c$, as computed in the preceding section and possible user-supplied contributions. The coefficients J and K are also user-supplied, so that neutral beam injection (J) and charge exchange (K) may be conveniently modeled.

2.4.1.7. Spatial discretization

Spatial derivatives are discretized as follows:

$$\frac{\partial}{\partial v}(Af)\bigg|_{i,j} \approx (A_{i,j+1}f_{i,j+1} - A_{i,j-1}f_{i,j-1})/2\Delta v_j, \qquad (2.4.28)$$

$$\frac{\partial}{\partial v}\left(B\frac{\partial f}{\partial v}\right)\bigg|_{i,j} \approx \left\{ B_{i,j+1/2}\left(\frac{f_{i,j+1} - f_{i,j}}{\Delta v_{j+1/2}}\right) \right. $$
$$\left. - B_{i,j-1/2}\left(\frac{f_{i,j} - f_{i,j-1}}{\Delta v_{j-1/2}}\right) \right\}\bigg/ \Delta v_j, \qquad (2.4.29)$$

$$\frac{\partial}{\partial v}\left(C\frac{\partial f}{\partial \theta}\right)\bigg|_{i,j} \approx \left\{ C_{i,j+1}\left(\frac{f_{i+1,j+1} - f_{i-1,j+1}}{2\Delta\theta_i}\right) \right. $$
$$\left. - C_{i,j-1}\left(\frac{f_{i+1,j-1} - f_{i-1,j-1}}{2\Delta\theta_i}\right) \right\}\bigg/ 2\Delta v_j, \qquad (2.4.30)$$

where

$$\Delta v_{j\pm1/2} = \pm(v_{j\pm1} - v_j), \qquad (2.4.31a)$$

$$\Delta v_j = \tfrac{1}{2}(\Delta v_{j-1/2} + \Delta v_{j+1/2}), \qquad (2.4.31b)$$

$$\Delta\theta_i = \tfrac{1}{2}(\theta_{i+1} - \theta_{i-1}), \qquad (2.4.31c)$$

$$B_{i,j\pm1/2} = \tfrac{1}{2}(B_{i,j} + B_{i,j\pm1}). \qquad (2.4.31d)$$

The terms $(\partial/\partial\theta)(DF)$, $(\partial/\partial\theta)(E(\partial f/\partial v))$ and $(\partial/\partial\theta)(F(\partial f/\partial\theta))$ are differenced analogously.

The boundary conditions must also be approximated; their discretization depends on the time-integration technique and will be discussed in a later section.

2.4.1.8. Time discretization

Equation (2.4.27) is time-integrated using either implicit operator splitting, an alternating-direction implicit (ADI) method, or fully implicit differencing.

Because implicit operator splitting and ADI are so similar, of the two only ADI will be described in detail. This is a two-step procedure, as follows:

$$\frac{f^{n+1/2} - f^n}{\Delta t/2} = \frac{1}{v^2}\frac{\delta}{\delta v}\left(A^n f^{n+1/2} + B^n\frac{\delta f^{n+1/2}}{\delta v} + C^n\frac{\delta f^n}{\delta\theta} \right)$$
$$+ \frac{1}{v^2\sin\theta}\frac{\delta}{\delta\theta}\left(D^n f^n + E^n\frac{\delta f^n}{\delta v} + F^n\frac{\delta f^n}{\delta\theta} \right) + K^n f^{n+1/2} + J^n, \qquad (2.4.32)$$

$$\frac{f^{n+1} - f^{n+1/2}}{\Delta t/2} = \frac{1}{v^2}\frac{\delta}{\delta v}\left(A^n f^{n+1/2} + B^n\frac{\delta f^{n+1/2}}{\delta v} + C^n\frac{\delta f^{n+1/2}}{\delta \theta}\right)$$

$$+ \frac{1}{v^2 \sin\theta}\frac{\delta}{\delta\theta}\left(D^n f^{n+1} + E^n\frac{\delta f^{n+1/2}}{\delta v} + F^n\frac{\delta f^{n+1}}{\delta\theta}\right) + K^n f^{n+1} + J^n, \qquad (2.4.33)$$

where the meshpoint indices i, j have been dropped, and where $\delta/\delta v$ and $\delta/\delta\theta$ denote the discretizations of (2.4.28)–(2.4.31). The mixed derivative terms are treated explicitly. This algorithm executes much faster than a fully implicit method since several small tridiagonal linear systems are inverted instead of a large nine-banded system. It is generally not necessary to iterate (2.4.32) and (2.4.33), since the Fokker–Planck coefficients depend on integrals of the distribution functions and therefore are slowly varying.

For full velocity space operation, (2.4.32) and (2.4.33), which hold at interior meshpoints, are joined by boundary value equations. Letting

$$G = Af + B\frac{\delta f}{\delta v} + C\frac{\delta f}{\delta\theta}, \qquad (2.4.34)$$

$$H = Df + E\frac{\delta f}{\delta v} + F\frac{\delta f}{\delta\theta}, \qquad (2.4.35)$$

Equation (2.4.32) is solved as follows:

At $\theta = 0$ ($i = 1$), $[1/(v^2 \sin\theta)](\delta H/\delta\theta)$ (the second term on the right-hand side) is replaced by

$$\frac{1}{v^2 \sin\theta}\frac{\delta H}{\delta\theta}\bigg|_{1,j} \approx \frac{2H^n_{3/2,j}}{v_j^2 \theta_{3/2}^2}. \qquad (2.4.36)$$

In cases where f is symmetric about $\pi/2$, the approximation

$$f^{n+1/2}_{I+1,j} = f^{n+1/2}_{I-1,j} \qquad (2.4.37)$$

is used, and in cases where $\theta_I = \pi$, $[1/(v^2 \sin\theta)](\delta H/\delta\theta)$ is replaced by

$$\frac{1}{v^2 \sin\theta}\frac{\delta H}{\delta\theta}\bigg|_{I,j} \approx -\frac{2H^n_{I-1/2,j}}{v_j^2 (\pi - \theta_{I-1/2})^2}. \qquad (2.4.38)$$

At $\theta = 0$ and π, the $C^n(\delta f^n/\delta\theta)$ term in (2.4.32) is set to 0.

At $v = 0$, the first term on the right-hand side of (2.4.32) is written as

$$\frac{1}{v^2}\frac{\delta G}{\delta v}\bigg|_{i,1} \approx \frac{3}{v_{3/2}^3}G^{n+1/2}_{i,3/2}, \qquad (2.4.39)$$

and

$$\frac{1}{v^2 \sin\theta}\frac{\delta H}{\delta\theta}\bigg|_{i,1} \approx 0, \qquad (2.4.40)$$

where

$$G^{n+1/2} = A^n f^{n+1/2} + B^n \frac{\delta f^{n+1/2}}{\delta v} + C^n \frac{\delta f^n}{\delta \theta}, \tag{2.4.41}$$

with the $C^n(\delta f^n/\delta\theta)$ term ignored at $\theta = 0$ and π.

Equation (2.4.33), the second half timestep, is solved analogously: At $\theta = 0$ $(i = 1)$, $[1/(v^2 \sin \theta)](\delta H/\delta\theta)$ is replaced by

$$\frac{1}{v^2 \sin \theta} \frac{\delta H}{\delta \theta} \approx \frac{2 H_{3/2, j}^{n+1}}{v_j^2 \theta_{3/2}^2}, \tag{2.4.42}$$

and at $\theta = \pi$ (when $\theta_I = \pi$),

$$\frac{1}{v^2 \sin \theta} \frac{\delta H}{\delta \theta} \approx - \frac{2 H_{I-1/2, j}^{n+1}}{v_j^2 (\pi - \theta_{I-1/2})^2}. \tag{2.4.43}$$

Here,

$$H^{n+1} = D^n f^{n+1} + E^n \frac{\delta f^{n+1/2}}{\delta v} + F^n \frac{\delta f^{n+1}}{\delta \theta}, \tag{2.4.44}$$

and the $C^n(\delta f/\delta\theta)$ term is again ignored. At $v = 0$,

$$f_{,1}^{n+1} = \sum_{i=1}^{I} b_i \left(f_{i,1}^{n+1/2} + \frac{\Delta t}{2} \frac{3}{v_{3/2}^3} \bar{G}_{i, 3/2}^{n+1/2} \right) \bigg/ \sum_{i=1}^{I} b_i, \tag{2.4.45}$$

where b_i is as in (2.4.10)–(2.4.12) and

$$\bar{G}^{n+1/2} = A^n f^{n+1/2} + B^n \frac{\delta f^{n+1/2}}{\delta v} + C^n \frac{\delta f^{n+1/2}}{\delta \theta}, \tag{2.4.46}$$

with the $\delta f/\delta\theta$ term ignored at $\theta = 0$ and π. This procedure assures that, in the absence of sources and losses, the particle density (2.4.13) does not change from timestep to timestep. The discrete boundary conditions for the fully implicit algorithm are similar, but actually much simpler since there is no time-splitting.

As mentioned in an earlier section, one is not restricted to full velocity space operation; that is, one may replace a subset of the above boundary conditions by Dirichlet conditions along given curves which are approximated by open polygons in discrete v, θ space.

The principal difference equations, (2.4.32) and (2.4.33), and the boundary conditions (2.4.36)–(2.4.46), may be cast in the following form:

$$-\alpha_{i,j}^n f_{i,j+1}^{n+1/2} + \beta_{i,j}^n f_{i,j}^{n+1/2} - \gamma_{i,j}^n f_{i,j-1}^{n+1/2} = \delta_{i,j}^n, \tag{2.4.47}$$

$$-\varepsilon_{i,j}^n f_{i+1,j}^{n+1} + \mu_{i,j}^n f_{i,j}^{n+1} - \nu_{i,j}^n f_{i-1,j}^{n+1} = \rho_{i,j}^n. \tag{2.4.48}$$

The procedure for solving (2.4.47) and (2.4.48) is the standard technique for solving tridiagonal systems (Richtmyer and Morton, 1967).

With fully implicit differencing one obtains a system of the form

$$Mf^{n+1} = y, \tag{2.4.49}$$

where f^{n+1}, the distribution function at the new timestep, is viewed as a vector with I times J components corresponding to meshpoints ordered rowwise or columnwise. The matrix M is a nine-banded matrix, a consequence of the nine-point difference algorithm. The discretization at the boundary points is similar to that for the ADI algorithm. The vector y depends on the distribution function at the old timestep, f^n, and the source term J. Both M and y are computed by FPPAC, and the crucial common blocks are compatible with those of the ILUCG package of Shestakov and Anderson (1983); hence implementation of ILUCG is, consequently, extremely simple.

2.4.1.9. *Energy transfer*

The rate of energy transfer to (or from) each general species is computed. The rate of energy density gained by species "a" due to collisions with species "b" is equal to the second moment of the operator which represents collisions between "a" and "b":

$$Q_{ab} = \pi m_a \int \int \left(\frac{\partial f_a}{\partial t} \right)^b v^4 \sin \theta \, dv \, d\theta \tag{2.4.50a}$$

$$= -2\pi m_a \Gamma_a \int \sin \theta \, d\theta \int G_a^b v \, dv, \tag{2.4.50b}$$

where G_a^b is that portion of G_a which depends on species "b"; i.e., in (2.1.10)–(2.1.11) the sum over "b" is replaced by "b" itself. (Note that the $\partial H_a/\partial \theta$ term in (2.1.14) does not contribute to energy transfer.) In the case where species "b" is isotropic, (2.4.50b) reduces to

$$Q_{ab} = -(4\pi)^2 m_a \Gamma_a \left(\frac{Z_b}{Z_a} \right)^2 \ln \Lambda_{ab}$$
$$\cdot \int v \, dv \cdot \left(\frac{m_a}{m_b} N_0(f_b) f_a + \frac{1}{3} \left(\frac{E_0(f_b)}{v} + v^2 M_0(f_b) \right) \frac{\partial f_a}{\partial v} \right), \tag{2.4.51}$$

and, furthermore, if f_b is Maxwellian with temperature T_b,

$$Q_{ab} = -(4\pi)^2 m_a \Gamma_a \left(\frac{Z_b}{Z_a} \right)^2 \ln \Lambda_{ab}$$
$$\cdot \int N_0(f_b) \left(\frac{m_a}{m_b} v f_a + \frac{T_b}{m_b} \frac{\partial f_a}{\partial v} \right) dv, \tag{2.4.52}$$

with $M_0(f_b)$, $N_0(f_b)$, and $E_0(f_b)$ as defined in (2.1.25), (2.1.26), and (2.1.28).

2.4.2. Applications

2.4.2.1. *Collisional loss of electrostatically confined species in a magnetic mirror*

In 1974 Pastukhov published expressions for the endloss rate of electrons in a magnetic square-well in the presence of a singly charged ionic species (Pastukhov, 1974). This formula was later extended to apply to either electrostatically confined ions or to electrons in a multispecies, multiply charged plasma (Cohen *et al.*, 1978). In this section results of the two-dimensional nonlinear Fokker–Planck equation are compared to that expression.

The equation

$$\frac{\partial f_a}{\partial t} = \left(\frac{\partial f_a}{\partial t}\right)_c + S_a + L_a \tag{2.4.53}$$

is solved, where a = ions or electrons, $(\partial f_a/\partial t)_c$ is given by (2.4.1), S_a is a low-energy source term and L_a is given by (2.1.66). A linearized Fokker–Planck equation is obtained by substituting a fixed Maxwellian of density n_a and temperature T_a for the distribution function f_a in (2.4.1). The validity of this approach is verified *a posteriori* through comparison with a nonlinear case.

The scaling of confinement time with potential is shown in Figs. 2.8 and 2.9.

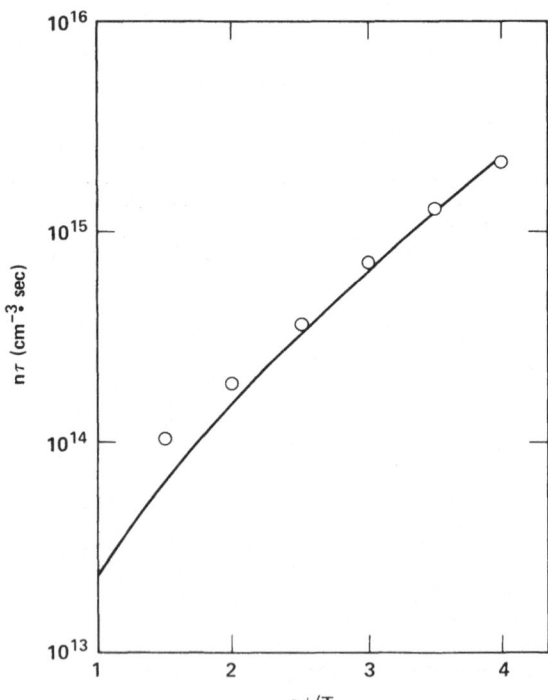

Figure 2.8. Ion confinement time versus potential for $R = 10$, $T = 30$ keV, $m = 2.5$ AMU, $Z = 1$. Code results are denoted by ◯; solid curve is generalized Pastukhov result.

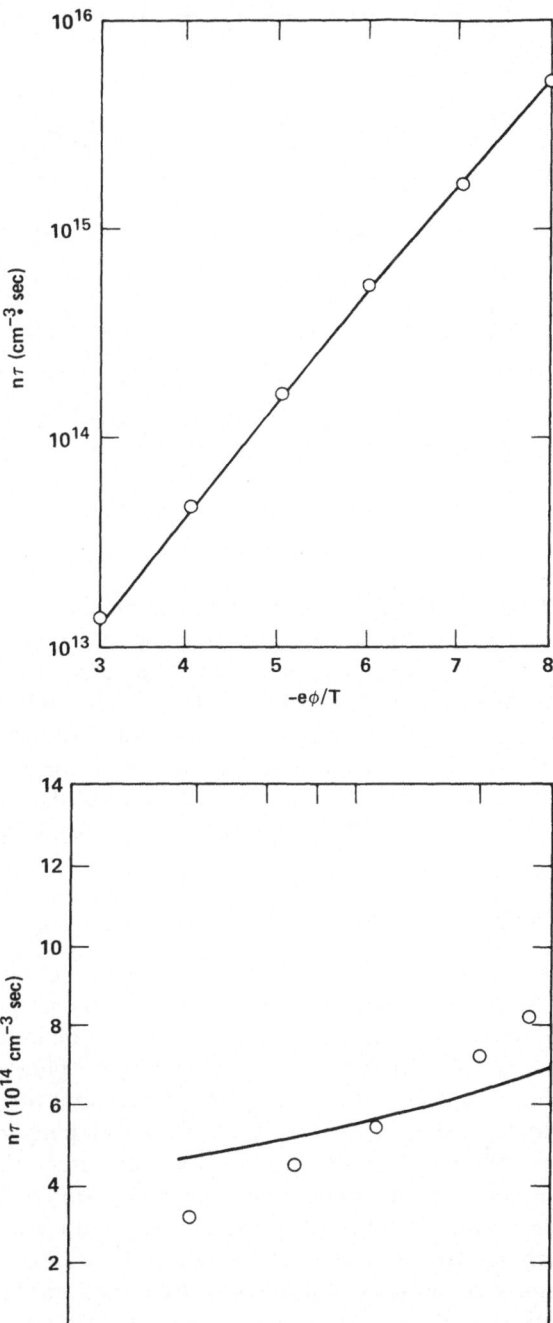

Figure 2.9. Electron confinement time versus potential for $R = 10$, $T = 45$ keV, $Z = 1$. Code results are denoted by ◯; solid curve is generalized Pastukhov result.

Figure 2.10. Ion confinement time versus mirror ratio for $e\phi/T = 3$, $T = 30$ keV, $m = 2.5$ AMU, $Z = 1$. Code results are denoted by ◯; solid curve is generalized Pastukhov result

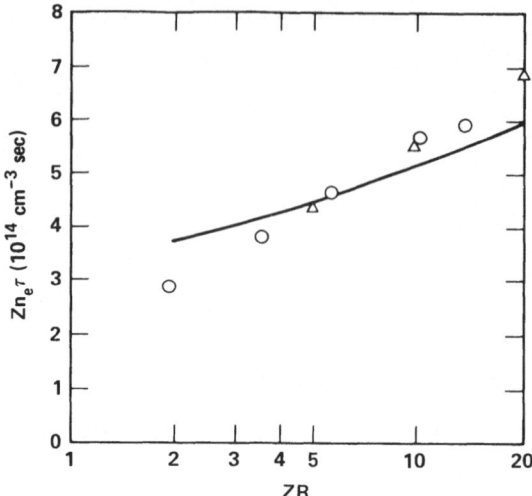

Figure 2.11. Electron confinement time versus mirror ratio for $-e\phi/\dot{T} = 6$, $T = 45$ keV. The multiplicative factor $\bar{Z} = \frac{1}{2}(1 + Z_i)$ is equal to unity for all code results except those denoted by $\triangle$ which have $(R = 5, Z = 1)$, $(R = 5, Z = 3)$, and $(R = 5, Z = 7)$. Code results are denoted by $\bigcirc$ and $\triangle$; solid curve is generalized Pastukhov result.

It is seen that there is excellent agreement between code and theory for $\bar{\phi} \equiv Ze\,\phi/T \gtrsim 2$; the scaling goes roughly as $\bar{\phi}\exp(\bar{\phi})$. Scaling with mirror ratio R is shown in Figs. 2.10 and 2.11. The generalized Pastukhov expressions vary more slowly with mirror ratio than do the Fokker–Planck results; the discrepancy is larger for ions than for electrons. Additionally, shown in Fig. 2.11 is the scaling of electron confinement time with the effective charge Z of the plasma. The Pastukhov prediction that the only Z-dependence is through the product ZR is verified by the numerical results. Further details may be found in Cohen *et al.* (1978).

2.4.2.2. *Ion confinement in tandem mirror plugs*

In a tandem mirror having supplemental electron heating or the capability of beam injection in the solenoid for two-component operation, the electron temperature T_e and plug potential ϕ_p can be independent of the plug injection energy $E_{\rm inj}$. In various applications, the plug-ion-particle lifetime $(n\tau)_p$ will vary from the electron-drag-dominated limit $(n\tau)_p \sim T_e^{3/2}$ to the ion-scattering-dominated limit $(n\tau)_p \sim \bar{E}_p^{3/2}$; the mean plug ion energy $\bar{E}_p$ is significantly less than $E_{\rm inj}$ in the former case and larger than $E_{\rm inj}$ in the latter case. A simple heuristic model of classical plug ion confinement in the general case is described. The model is useful for reactor studies because it allows one to replace two-dimensional, nonlinear Fokker–Planck computations by a few simple equations describing particle and energy confinement in a magnetic mirror.

When both angle scattering and drag are important, we write $(n\tau)_p$ as the sum of the reciprocals of the scattering and drag times

$$(n\tau)_p = \left\{\left[\left(\frac{C_1}{\ln \Lambda_{ii}}\right)\left(\frac{M_i}{M_D}\right)^{1/2} E_{inj}^{3/2} \log_{10}(R_{eff})\right]^{-1}\right.$$
$$\left. + \left[\left(\frac{2 \times 10^{13}}{\ln \Lambda_{ei}}\right)\left(\frac{M_i}{M_D}\right) T_e^{3/2} \ln\left(\frac{E_{inj}}{\bar{E}_L}\right)\right]^{-1}\right\}^{-1},$$

(2.4.54)

where C_1 is an adjustable constant to be fit to Fokker–Planck calculations, (M_i/M_D) is the mass of $Z = 1$ ions normalized to that of deuterium, R_{eff} is an effective mirror ratio defined in terms of the actual magnetic mirror ratio R_m and the positive ambipolar potential ϕ_p as seen by the plug ions

$$R_{eff} = \frac{R_m \sin^2 \theta_{inj}}{1 + (\phi_p/E_{inj})}$$

(2.4.55)

and $\bar{E}_L$ is the mean energy of the escaping ions. When T_e is sufficiently large compared to $\bar{E}_{inj}$, $(n\tau)_p$ is reduced to scattering-dominated loss (the first term in (2.4.54)); when T_e is small enough compared to $\bar{E}_{inj}$, $(n\tau)_p$ is reduced to drag-dominated loss (second term in (2.4.54)). When the injection point in velocity space approaches the loss boundary in the presence of an ambipolar potential, $R_{eff} \to 1$ and $(n\tau)_p \to 0$, as must be the case. The factor $\ln(E_{inj}/\bar{E}_L)$ in the second term of (2.4.54) gives the number of e-folding drag times for an ion injected with energy E_{inj} to be lost. When no energy is exchanged between ions and electrons (limit of high T_e), energy conservation requires $\bar{E}_L \to E_{inj}$, even though upward scattering will give $\bar{E}_p > E_{inj}$. In the drag-dominated limit, ions will escape with the cut-off energy,

$$E_c = \frac{\phi_p}{R_m \sin^2 \theta_{inj} - 1}.$$

(2.4.56)

In the general case, we write

$$\bar{E}_L = \frac{E_{inj}}{1 + (\tau_{ii}/\tau_{drag})} + \frac{E_c(\tau_{ii}/\tau_{drag})}{1 + (\tau_{ii}/\tau_{drag})},$$

(2.4.57)

where

$$\left(\frac{\tau_{ii}}{\tau_{drag}}\right) = C_2\left(\frac{E_{inj}}{T_e}\right)^{3/2} \frac{\log_{10}(R_{eff})}{\ln(E_{inj}/\bar{E}_L)}\left(\frac{\ln \Lambda_{ei}}{\ln \Lambda_{ii}}\right)\left(\frac{M_D}{M_i}\right)^{1/2},$$

(2.4.58)

with C_2 a second adjustable constant to be determined by Fokker–Planck calculations. (In this section, all energies are in keV, and n is in cm^{-3}.)

The mean energy of the confined plug ions is determined by ion energy balance in the plugs

$$\frac{n_p^2 E_{inj}}{(n\tau)_p} = \frac{n_p^2}{(n\tau)_{drag}}\bar{E}_p + \frac{n_p^2}{(n\tau)_p}\bar{E}_L,$$

(2.4.59)

where

$$(n\tau)_{drag} = 2 \times 10^{13}(M_i/M_D)T_e^{3/2}/\ln \Lambda_{ei}.$$

(2.4.60)

Table 2.3. Test case data and model parameters derived from Fokker–Planck code results.

Case	R_m	T_e (keV)	ϕ_p (keV)	E_{inj} (keV)	θ_{inj} (degrees)	C_1 (10^{12} keV$^{-3/2}$ cm^{-3} s)	C_2 ($\times 10^{-2}$)
1	3	50	300	804	90	3.13	11.2
2	3	50	300	804	75	3.42	10.6
3	3	50	300	804	60	4.96	7.46
4	2	30	120	1180	90	6.31	8.48
5	2	30	240	1180	90	3.70	8.89
6	2	70	280	1180	90	4.35	8.85
7	2	70	560	1180	90	3.44	10.3
8	2	30	120	399	90	4.35	9.10
9	2	30	240	399	90	3.55	9.92
10	2	70	280	399	90	5.30	9.09
11	4	30	120	1180	90	5.02	9.91
12	4	30	240	1180	90	2.78	10.8
13	4	70	280	1180	90	3.19	12.3
14	4	70	560	1180	90	2.58	12.9
15	4	30	120	399	90	3.25	12.6
16	4	30	240	399	90	2.76	14.5
17	4	70	280	399	90	4.09	14.8

The adjustable constants C_1 and C_2 in (2.4.54) and (2.4.58) are determined by solving (2.4.53), in which mirror ratio, electron temperature, potential, injection energy, and injection angle are varied independently. Results from a number of representative test cases are given in Table 2.3. The average values of C_1 and C_2 derived from this table are

$$C_1 = 3.9 \times 10^{12} \text{ keV}^{-3/2} \text{ cm}^{-3} \text{ s}, \qquad (2.4.61)$$

$$C_2 = 1.1 \times 10^{-1}. \qquad (2.4.62)$$

The ion–particle lifetime $(n\tau)_p$ and the mean energy of the escaping ions $\bar{E}_L$ can be estimated for any set of parameters by using the model given by (2.4.54)–(2.4.60). Note that (2.4.57) is nonlinear in $\bar{E}_L$, so that an iterative solution is required. Comparisons between the model predictions and Fokker–Planck code results are given in Table 2.4. The agreement is generally within $\pm 5\%$ over the range of parameters covered by this study. This is nearly equal to the uncertainty in the Fokker–Planck results arising from the use of a finite grid in velocity space. We conclude that, to describe classical confinement in the plugs of a tandem reactor in which charge exchange is small compared to ionization, the model given here can be used in lieu of Fokker–Planck calculations for the range of parameters tested. By including charge exchange off the beam, future work on the model may extend its usefulness to describe classical confinement in mirror experiments such as the TMX plugs.

Table 2.4. Comparison of Fokker–Planck code and model results for particle confinement and loss energy.

Case	Particle confinement, $(n\tau)_p$ $(10^{13}$ cm^{-3} s)			Loss energy, E_L (keV)		
	Code	Model	% Difference	Code	Model	% Difference
1	24.7	25.8	+4.5	346	348	+0.6
2	23.3	24.1	+3.4	368	365	−0.8
3	17.4	18.5	+6.3	465	431	−7.3
4	20.5	21.0	+2.4	354	321	−9.3
5	16.2	17.1	+5.6	435	412	−5.3
6	31.8	33.0	+3.8	625	595	−4.8
7	19.5	20.3	+4.1	780	774	−0.8
8	6.89	7.02	+1.9	241	233	−3.3
9	3.38	3.53	+4.4	306	304	−0.7
10	4.59	4.24	−7.6	354	354	+0.0
11	30.6	31.0	+1.0	200	190	−5.0
12	27.3	28.2	+3.3	224	223	−0.4
13	58.8	59.2	+0.7	366	382	+4.4
14	48.0	50.3	+4.8	430	451	+4.9
15	13.8	13.8	+0.0	146	153	+4.8
16	11.1	11.3	+1.8	168	181	+7.7
17	20.2	18.5	−8.4	235	249	+6.0

2.4.2.3. *Distortion of bulk ions from Maxwellians in the tokamak fusion test reactor* (TFTR)

It has been shown that during the heating of a warm-background ion species by a more energetic species, the warm ion distribution function departs from that of a Maxwellian (Cordey, 1975). It has also been shown that the velocity dependence of the particle and energy loss mechanisms is a crucial factor in determining the manner in which the bulk ions deviate from Maxwellians (Bittoni et al., 1980). The question of the extent to which these distortions cause an enhanced fusion reactivity in a TFTR-like plasma is addressed.

In particular, a plasma composed of fixed electrons, bulk deuterons, bulk tritons, fixed hot deuterons and fixed hot α-particles is considered. The parameters are chosen from on-axis results of Mirin and Jassby (1980). The hot deuterons arise from neutral-beam injection, and bulk α-particles are ignored. The electrons are taken to be Maxwellian at a density of 9.74×10^{13} cm^{-3} and a temperature of 9 keV. The bulk deuterons and tritons have densities of 5×10^{13} cm^{-3} and 3×10^{13} cm^{-3}, respectively. The hot deuterons are computed by using Cordey's (1975) solution for fast ions; their mean energy turns out to be 71.0 keV, and their density is taken as 1.5×10^{13} cm^{-3}. The hot alphas are computed similarly; their mean energy is 1129 keV and their density is 1.2×10^{12} cm^{-3}.

Table 2.5. (σv) comparison for $T_e = 9$ keV and $\bar{E}_{i,\text{FP}} = 39$ keV.

β	$\tau_0(s)$	$\overline{\sigma v}_{\text{FP}}(\times 10^{-16}\ \text{cm}^3\ \text{s}^{-1})$	$\bar{E}_{i,\text{max}}(\text{keV})$	$\overline{\sigma v}_{\text{max}}(\times 10^{-6}\ \text{cm}^3\ \text{s}^{-1})$
-2	22.5	5.38	38.7	5.23
-1	4.80	5.37	38.9	5.26
0	1.11	5.35	39.1	5.28
1	0.286	5.33	39.1	5.29
2	0.0828	5.31	39.0	5.27
3	0.0306	5.30	39.1	5.29
4	0.0140	5.29	39.2	5.30

In Scenario I, Fokker–Planck equations for the bulk deuteron and triton distribution functions are solved, subject to collisions with all charged species and with an energy loss term, i.e.,

$$\frac{\partial f}{\partial t} = \left(\frac{\partial f}{\partial t}\right)_c + \frac{1}{v^2}\frac{\partial}{\partial v}\left(\frac{v^3}{2\tau(v)}f\right), \tag{2.4.63}$$

where $\tau(v) = \tau_0[\max(v, v_0)]^\beta$. The quantity τ_0 is first chosen to give a steady-state mean bulk ion energy, $\bar{E}_{i,\text{FP}} = (n_D E_D + n_T E_T)/(n_D + n_T)$, of 39 keV, and v_0 corresponds to 3 keV. The parameter β is freely varied to represent different loss mechanisms, such as transport or charge exchange. Equation (2.4.63) is integrated to steady state. In Scenario II, the bulk deuteron and triton distribution functions are assumed to be Maxwellian, and (2.4.63) is integrated over velocity space to yield rate equations for their mean energies. These rate equations are then integrated to steady state.

Comparisons for $\beta = -2, -1, 0, 1, 2, 3$, and 4 are carried out. Results are displayed in Table 2.5. The third and fifth columns compare the thermal fusion reaction rates (computed using the same technique as in Marx et al. (1976)) of the two scenarios, and the fourth column contains the mean bulk ion energy for Scenario II. It is readily concluded that there is substantially no difference in the fusion reactivity between the two scenarios.

These cases are all rerun at lower energy where it is hoped that "tail-pulling"

Table 2.6. (σv) comparison for $T_e = 6.67$ keV and $\bar{E}_{i,\text{FP}} = 15$ keV.

β	$\tau_0(s)$	$\overline{\sigma v}_{\text{FP}}(\times 10^{-17}\ \text{cm}^3\ \text{s}^{-1})$	$\bar{E}_{i,\text{max}}(\text{keV})$	$\overline{\sigma v}_{\text{max}}(\times 10^{-17}\ \text{cm}^3\ \text{s}^{-1})$
-2	0.842	8.99	14.6	8.91
-1	0.292	9.30	14.8	9.28
0	0.110	9.58	15.0	9.65
1	0.0445	9.83	14.9	9.52
2	0.0211	10.2	15.1	9.75
3	0.0117	10.4	16.1	11.4
4	0.00794	10.6	20.5	20.0

Figure 2.12. Temperature ratio
and partial density versus
energy for Scenario I, $\beta = 4$
case having mean energy of 15
keV. T = local temperature
based on slope of distribution
function, T_0 = average temperature, n = density of particles having energy $\leq E$, and
n_0 = total density.

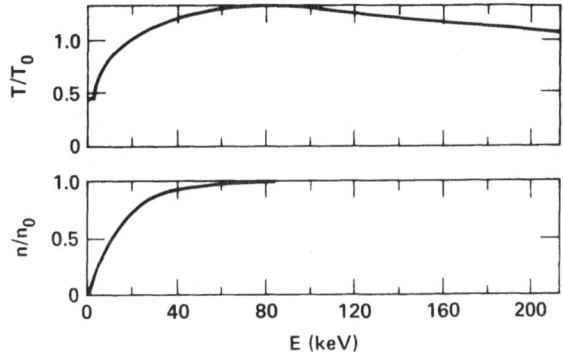

will prove significant. In particular, T_e is lowered to 6.67 keV and the target
ion energy $\bar{E}_{i,\mathrm{FP}}$ is lowered to 15 keV. Results are displayed in Table 2.6.
The pheonomenon of tail-pulling is displayed in Fig. 2.12. For $\beta \leq 2$, there is
very little difference in fusion reactivity. For $\beta = 3$, a 10% higher reactivity is
predicted for Scenario II, and for $\beta = 4$ the predicted reactivity for Scenario II
is almost twice that for Scenario I.

To understand this startling result, the energy equation must be examined
in detail. For a Maxwellian at temperature T_0, the integrated form of (2.4.63)
may be written

$$\frac{dT_0}{dt} = \left.\frac{dT_0}{dt}\right|_{\mathrm{coll.}} + \left.\frac{dT_0}{dt}\right|_{\mathrm{loss}}, \tag{2.4.64}$$

where the first term on the right-hand side of (2.4.64) represents energy gain
(or loss) due to collisions with the other charged species, and the second
term represents noncollisional energy loss. The collisional term is naturally a
decreasing function of T_0. However, for small v_0, the noncollisional energy loss
term (equal to $\int_0^\infty f(v)\tau^{-1}(v)v^4\,dv$) goes as $-T_0^{(1-\beta/2)}$, and for $\beta > 2$ it is also
(ignoring sign) a decreasing function of T_0. Thus, as β increases beyond 2, the
right-hand side of (2.4.64) becomes a relatively insensitive function of T_0. In
the case where the bulk ion distribution functions are not Maxwellian, the
same type of behavior holds, and for large β the precise rate at which energy
is gained or lost depends much more strongly on the actual shape of the
distribution function. For $\beta = 4$ the insensitivity with respect to T_0, combined
with differences in distribution function shape, results in a mean bulk ion
energy of 20.5 keV for Scenario II versus 15 keV for Scenario I, and it is this
energy difference which is responsible for the larger fusion reactivity. It should
be pointed out that for this case there is a lot of tail-pulling (see Fig. 2.12), some
of which is brought about by the energy dependence of the loss term—but it
is that very loss term which causes a much higher fusion reactivity in Scenario
II, which more than offsets the increase in fusion reactivity in Scenario I due
to the tail-pulling.

For $n_\alpha = 0$ and $\beta = 0$, results of our model may be compared with those of Bittoni *et al.* (1980), in which an analytic prediction for the temperature ratio T/T_0 at large energy is given. Using Eq. (9) of Bittoni *et al.* (1980) and the notation of that reference, and noting that $m_f = m_i$, $Z_f = 1$ and $\tau_{cx}^{-1} = 0$, the temperature ratio T/T_0 may be expressed as

$$\frac{T_{\text{tail}}}{T_{\text{act}}} = \frac{1 + \dfrac{2n_f \bar{\varepsilon}_f}{3n_i^* T_{\text{act}}} + \left(\dfrac{v}{v_c}\right)^3 \dfrac{n_e T_e}{n_i^* T_{\text{act}}}}{1 + \dfrac{n_f}{n_i^*} + \left(\dfrac{v}{v_c}\right)^3 \dfrac{n_e}{n_i^*}\left(1 + \dfrac{\tau_s}{2\tau_e}\right)}. \tag{2.4.65}$$

For $T_e = 6.67$ keV and $\bar{E}_{i,\text{FP}} = 15$ keV, the resulting deuteron energy E_D is 15.2 keV and the resulting τ_0 is 0.131 s. Substituting these values along with $n_i^* = 7.0 \times 10^{13}$ cm^{-3}, $n_f = 1.5 \times 10^{13}$ cm^{-3}, $n_e = 9.5 \times 10^{13}$ cm^{-3} and $\bar{\varepsilon}_f = 71.0$ keV, (2.4.65) for the deuterons becomes

$$T/T_0 = \frac{1.65 + 5.33 \times 10^{-4} E^{3/2}}{1 + 1.93 \times 10^{-3} E^{3/2}}. \tag{2.4.66}$$

A comparison is shown in Table 2.7. It can be seen that there is, as expected, excellent agreement for $E \gg \bar{E}_{i,\text{FP}}$ and rather poor agreement at lower energies.

The principal point that has been addressed is the accuracy of the predicted fusion reactivity under the assumption of a Maxwellian background plasma. For the above cases that prediction, for the most part, turns out to be quite accurate.

2.4.2.4. *Electrostatically trapped electrons in thermal barrier tandem mirrors*

Analytic expressions for the density and energy of electrostatically trapped electrons in the thermal barrier of a tandem mirror, valid for small to intermediate mirror ratios ($0 \leq R \leq 2$) with and without passing particles, have been derived (Rensink *et al.*, 1984). Results of the two-dimensional nonlinear Fokker–Planck equation are compared to those expressions. Equation (2.4.53) is solved for the distribution function of trapped electrons f_e subject to the boundary condition $f_e = f_u$ on the trapped/passing separatrix (see Section

Table 2.7. Temperature ratio T/T_0 versus energy.

E (keV)	T/T_0 (Bittoni)	T/T_0 (Mirin)
20	1.45	1.0
60	1.00	0.9
120	0.66	0.7
140	0.60	0.6

Figure 2.13. Density of trapped particles (n_{tr}) versus mirror ratio (R).

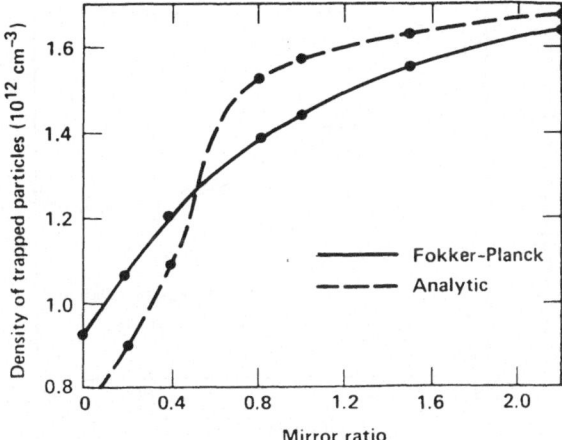

2.1.3), where f_u is the Maxwellian distribution of passing electrons at density n_u and temperature T_u.

This comparison is first carried out for no passing particles ($n_u = 0$). The confining potential Φ is 0.23 keV and electrons at an energy of 1 eV are injected at a rate of 1×10^{16} cm^{-3} s^{-1}. An equal number of ions are present with mean energy E_i. Both E_i and the mirror ratio R are varied.

Figures 2.13 and 2.14 show the density and mean energy, respectively, of trapped particles versus mirror ratio for both the numerical and analytic models. For these cases, $E_i \equiv 80$ keV. As R increases, the density of trapped particles n_{tr} increases owing to better confinement and their mean energy E_{tr} increases since the trapped particles encompass a wider range of velocity

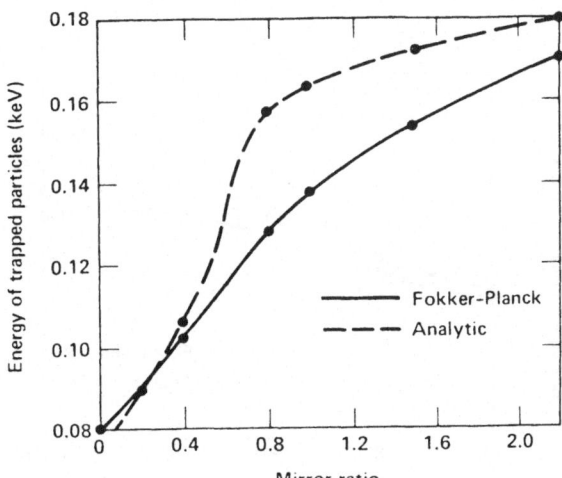

Figure 2.14. Mean energy of trapped particles ($E_{tr} \equiv \frac{3}{2}T_{tr}$) versus mirror ratio ($R$).

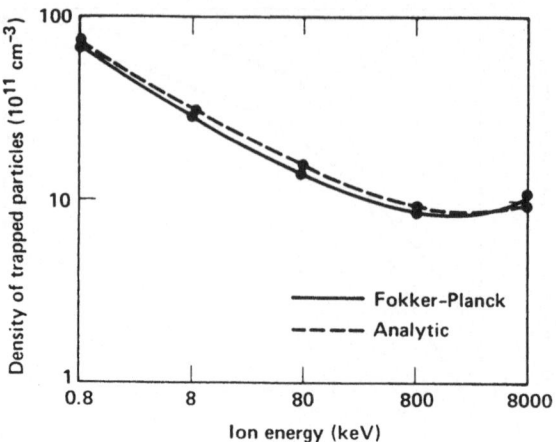

Figure 2.15. Density of trapped particles (n_{tr}) versus ion energy (E_i).

space. The code and analytic results agree to within 10% for density and 20% for energy.

Figures 2.15 and 2.16 show the density and energy of trapped particles versus ion energy at fixed mirror ratio $R = 0.8$ for both the numerical and analytic models. As E_i increases, T_{tr} increases because of the increasing energy transfer rate. The density n_{tr} correspondingly decreases as more particles are diffused across the separatrix. This trend continues until the ion energy contribution to the collision time outweighs the difference in electron and ion energies. The code results and analytic predictions for density agree to within 10%. The energy agreement is excellent at low energies and drops off at higher energies because the ion contribution to the electron energy diffusion operator has been neglected in obtaining the analytic expressions for the fluxes.

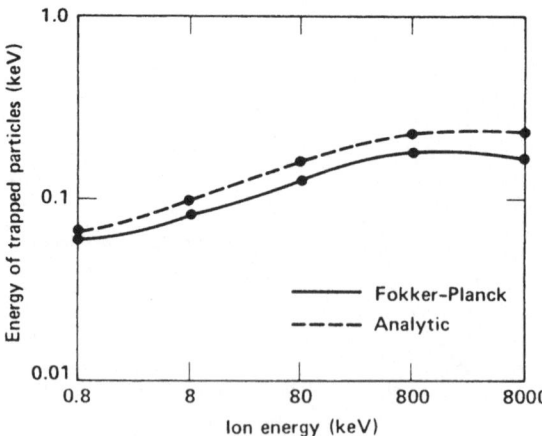

Figure 2.16. Mean energy of trapped particles (E_{tr}) versus ion energy (E_i).

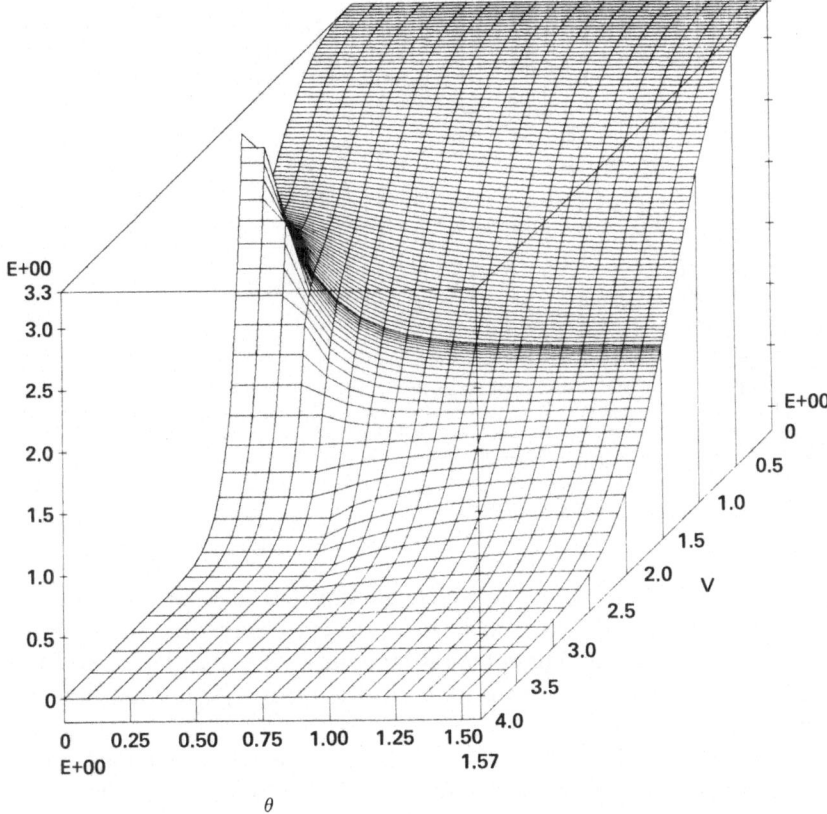

Figure 2.17. Electron distribution function for case with passing particles.

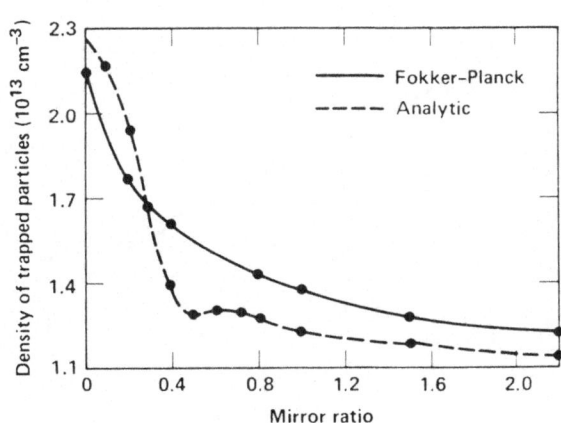

Figure 2.18. Density of trapped particles versus mirror ratio (with passing particles).

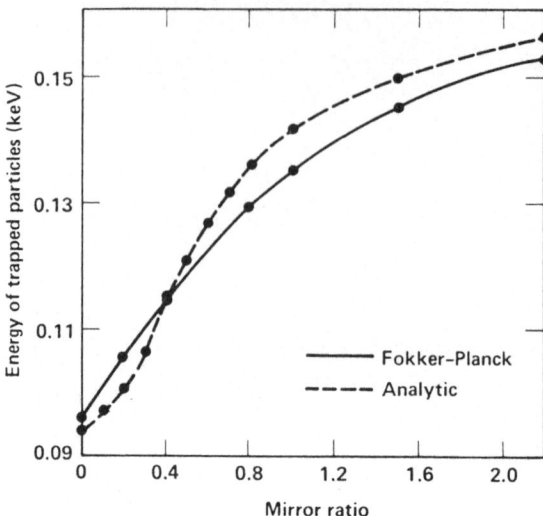

Figure 2.19. Mean energy of trapped particles versus mirror ratio (with passing particles).

Cases are now considered in which passing particles are present. Their density at the mirror throat is $n_m = 1 \times 10^{12}$ cm^{-3} and their temperature is $T_u = 0.06$ keV. The confining potential remains at $\Phi = 0.23$ keV, and there is no additional source of electrons. A typical electron distribution function is shown in Fig. 2.17.

Figures 2.18 and 2.19 show the density and energy of trapped particles versus mirror ratio (at $E_i \equiv 8$ keV), and Figs. 2.20 and 2.21 show the variation with ion energy (at $R \equiv 2.2$). Note that density is now a decreasing function of mirror ratio, presumably because, as the separatrix moves out in velocity space, the relative trapping rate of passing particles decreases. There is a kink

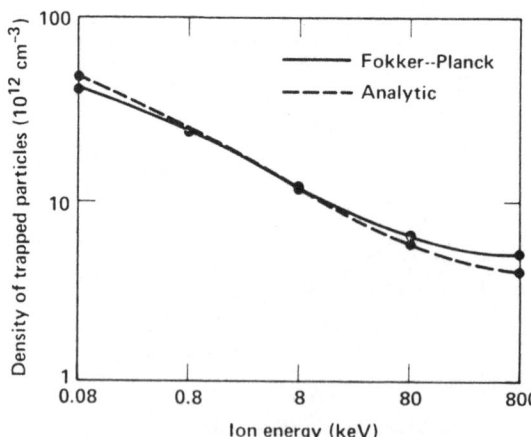

Figure 2.20. Density of trapped particles versus ion energy (with passing particles).

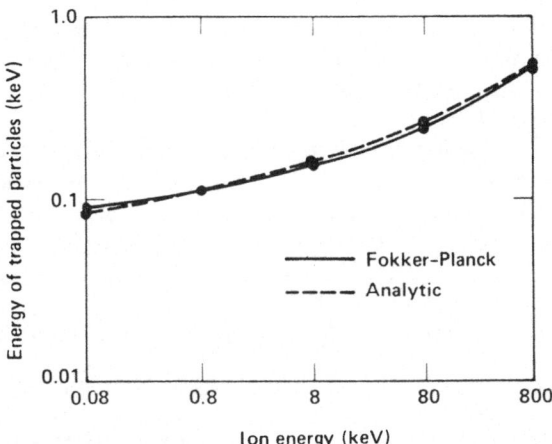

Figure 2.21. Mean energy of trapped particles versus ion energy (with passing particles).

in the analytically computed n_{tr}-versus-R curve (Fig. 2.18) as a result of the model distribution function chosen to relate n_{tr} to the density of the zeroth order Maxwellian, n_t. An n_t-versus-R plot does not have such a kink (see Rensink *et al.*, 1984). The Fokker–Planck and analytic models generally agree to within 10%, except at intermediate mirror ratios where the agreement is not quite as good.

Further details may be found in Rensink *et al.* (1984).

References

D. J. BenDaniel, *Plasma Phys.*, **3**, 235 (1961).

H. L. Berk, H. P. Furth, D. L. Jassby, R. M. Kulsrud, C. S. Liu, M. N. Rosenbluth, P. H. Rutherford, F. H. Tenney, T. Johnson, J. Killeen, A. A. Mirin, M. E. Rensink, and C. W. Horton, Jr., Plasma Physics and Controlled Nuclear Fusion Research, 1974 (IAEA, Vienna) III, 569 (1975).

E. Bittoni, J. G. Cordey, and M. Cox, *Nucl. Fusion*, **20**, 931 (1980).

R. H. Cohen, M. E. Rensink, T. A. Cutler, and A. A. Mirin, *Nucl. Fusion*, **18**, 1229 (1978).

J. G. Cordey, Plasma Physics and Controlled Nuclear Fusion Research, Tokyo, 1974 (IAEA, Vienna) **1**, 623 (1975).

J. G. Cordey, K. D. Marx, M. G. McCoy, A. A. Mirin, and M. E. Rensink, *J. Comput. Phys.*, **28**, 115 (1978).

J. G. Cordey and C. J. H. Watson, Controlled Fusion and Plasma Physics, (Proceedings of the Fifth European Conference, Grenoble, 1972) **1**, 98 (1972).

J. G. Cordey and C. J. H. Watson, Plasma Physics and Controlled Nuclear Fusion Research, Tokyo, 1974 (IAEA, Vienna) **2**, 643 (1975).

J. M. Dawson, H. P. Furth, and F. H. Tenney, *Phys. Rev. Lett.*, **26**, 1156 (1971).

T. K. Fowler and M. Rankin, *Plasma Phys.*, **4**, 311 (1962).

T. K. Fowler and M. Rankin, *Plasma Phys.*, **8**, 121 (1966).

H. P. Furth and S. Yoshikawa, *Phys. Fluids*, **13**, 2593 (1970).

A. H. Futch, Jr., J. P. Holdren, J. Killeen, and A. A. Mirin, *Plasma Phys.*, **14**, 211 (1972).

C. Hastings, Jr., *Approximations for Digital Computers*, Princeton University Press, Princeton, NJ, 1955.

A. N. Kaufman, US Atomic Energy Commission, Doc. TID-7520, Part 2, 387 (1956).

J. Killeen, W. Heckrotte, and G. Boer, *Nucl. Fusion*, Supplement, Part 1, 183 (1962).

J. Killeen and A. H. Futch, Jr., *J. Comput. Phys.*, **2**, 236 (1968).

J. Killeen and K. D. Marx, *Meth. Comput. Phys.*, **9**, 422 (1970).

J. Killeen and A. A. Mirin, Fourth Conference on Numerical Simulation of Plasmas, NRL, Washington, DC, 685 (1970).

J. Killeen and A. A. Mirin, College in Theoretical and Computational Plasma Physics, Trieste (IAEA, Vienna, 1978), p. 27.

K. D. Marx, A. A. Mirin, M. G. McCoy, M. E. Rensink, and J. Killeen, *Nucl. Fusion*, **16**, 702 (1976).

M. G. McCoy, A. A. Mirin and J. Killeen, "A Vectorized Fokker–Planck Package for the Cray-1," Report UCRL-83206, Scientific Computer Exchange Meeting, Livermore, CA (1979).

M. G. McCoy, A. A. Mirin, and J. Killeen, *Comput. Phys. Commun.*, **24**, 37 (1981).

A. A. Mirin, "Isotions, A One-Dimensional Multispecies Fokker–Planck Computer Code," Lawrence Livermore National Laboratory Report UCRL-51616 (1974).

A. A. Mirin, J. Killeen, and C. J. H. Watson, *Nucl. Fusion*, **17**, 47 (1977).

A. A. Mirin and D. L. Jassby, *IEEE Trans. Plasma Sci.*, **PS-8**, 503 (1980).

V. P. Pastukhov, *Nucl. Fusion*, **14**, 3 (1974).

R. F. Post, *Phys. Fluids*, **4**, 902 (1961).

R. F. Post, T. K. Fowler, J. Killeen, and A. A. Mirin, *Phys. Rev. Lett.*, **31**, 280 (1973).

M. E. Rensink, R. H. Cohen, A. A. Mirin, and G. P. Tomaschke, *Nucl. Fusion*, **24**, 49 (1984).

R. D. Richtmyer and K. W. Morton, *Difference Methods for Initial Value Problems*, Wiley (Interscience), New York, 1967.

M. N. Rosenbluth, W. M. MacDonald, and D. L. Judd, *Phys. Rev.*, **107**, 1 (1957).

A. I. Shestakov and D. V. Anderson, *Comput. Phys. Commun.*, **30**, 31 (1983).

L. Spitzer, Jr., *Physics of Fully Ionized Gases*, 2nd ed., Wiley (Interscience), New York, 1962.

E. E. Yushmanov, *Zh. Eksp. Teor, Fiz.*, **49**, 588 (1966) and *Sov. Phys., JETP*, **22**, 409 (1966).

CHAPTER 3

Collisional Kinetic Models of Multispecies Plasmas in Nonuniform Magnetic Fields

3.1. Mathematical Model

It is generally the case in magnetic fusion devices that the magnetic field is nonuniform, but varies weakly on the scale of the gyro-motion, the fastest nearly recurrent motion of the charged particles comprising the plasma. However, motion along the direction of the field inexorably carries the particle through finite variations of the field, and the environment in which the charged particle is immersed as it executes its motion may vary significantly with these field variations. It is often necessary to consider this nonuniformity in order to involve salient features distinctive of a particular device in phenomena in which they play a significant role.

An example of such a feature is the presence of trapped particles in tokamaks. Guiding center orbits which are nearly trapped and oppositely directed may be topologically adjacent in a tokamak and thus may populate one another through diffusion (collisional or otherwise). In a uniform magnetic field calculation this could not occur. The only recurrent motion in that case is the gyro-motion and oppositely directed orbits in the preceding sense are inaccessible to one another through small angle collisions or other similarly diffusive processes.

The electrical conductivity in the direction of the magnetic field in a tokamak is affected by the presence of a population of trapped electrons as was shown by Hinton and Hazeltine (1976) and Connor *et al.* (1973). Since most trapped particles execute nearly recurrent "bounce" orbits along the direction of the ohmic electric field, the effect of the field on them is effectively nullified, averaged away by their motion. Trapped particles thus carry no electrical current and cannot contribute to the conductivity.

The use of cyclotron resonance heating in fusion devices and the production of plasma currents, to allow steady-state operation in tokamaks by the application of radio frequency (RF) microwave fields, is currently of great interest. The nonuniformity of the magnetic field bears upon each of these phenomena.

One of the earliest evaluations of ion cyclotron resonance heating (ICRH) as a reliable mechanism for auxiliary heating from the standpoint of Fokker–Planck theory was done by Stix (1975). Although the groundwork for this approach had been laid earlier by Rosenbluth *et al.* (1957) and Kennel and Englemann (1966) who developed the uniform field, infinite homogeneous media theory, a unified theory of particle–wave interactions had yet to be applied. The Stix approach was generalized for reactor studies by several authors, notably Kesner (1978), Scharer *et al.* (1979), Blackfield and Scharer (1982), and more recently Hwang *et al.* (1983). Each of these approaches is a variant of the uniform field infinite media calculation, insofar as the kinetic theory is concerned, with various finite geometry (toroidal or mirror) corrections. More recently, Mauel (1982) has modernized the quasilinear approach following the example of Berk (1978) and Bernstein and Baxter (1981).

Among the first to explore the possibilities of steady-state operation of tokamaks with RF current drive were Fisch and Bers (1978); variations on this theme have been provided more recently by Fisch (1980), Fisch and Karney (1981), Harvey *et al.* (1981), and others.

In each case discussed, the interaction between particles and waves is localized by a resonance condition involving the particle motion, the fluctuating microwave field, and the steady magnetic field. The strength of the interaction is determined by the time during which the particle motion and the fluctuating field are correlated. In a uniform field, decorrelation is effected by a finite width spectrum of waves, interference among members of the wave ensemble decorrelating the interaction of the particle with any one single member. In a nonuniform field, the resonance is localized in physical space also through the dependence on the steady magnetic field. The particle motion itself thus provides a decorrelation mechanism which in many cases is overriding. Moreover, this provides a natural mechanism whereby a much broader population of orbits interact resonantly with the wave field.

3.1.1. Bounce-averaged Fokker–Planck theory

Consider a multispecies magnetized plasma whose evolution is governed by the Fokker–Planck equation in which the dominant diffusive mechanisms are small angle Coulomb collisions, wave–particle interactions within the purview of weak turbulence theory, and a small amplitude steady electric field

$$\frac{\partial f}{\partial t} + \mathbf{v} \cdot \nabla f + \nabla_{\mathbf{v}} \cdot \left(\frac{q}{m} \left(\mathbf{E} + \frac{\mathbf{v}}{c} \times \mathbf{B} \right) f + \Gamma_c \right) = 0. \tag{3.1.1}$$

Here Γ_c represents the current density in velocity space due to small angle Coulomb collisions as first shown by Rosenbluth *et al.* (1957). For collisions of all species j on species i, Γ_c may be written in the concise (Landau) form

$$\Gamma_{ci} = \sum_i 2\pi \frac{q_i^2 q_j^2}{m_i} \ln \Lambda_{ij} \int d^3 v_j \, \mathsf{M} \cdot \left(\frac{1}{m_i} \nabla_{\mathbf{v}_i} - \frac{1}{m_j} \nabla_{\mathbf{v}_j} \right) f_i(\mathbf{r}, \mathbf{v}_i, t) f_j(\mathbf{r}, \mathbf{v}_j, t), \tag{3.1.2}$$

where

$$M = \nabla_{v_i} \nabla_{v_j} g = \frac{1}{g^3}(g^2 1 - gg); \qquad g = |g| = |v_i - v_j|. \qquad (3.1.3)$$

The additional complication of the field nonuniformity appears as a nontrivial coupling between the configuration and velocity space arguments of the distribution functions f_i, f_j. The electric and magnetic fields in (3.1.1), E and B, are composed of a fluctuating component arising principally due to the presence of antennae at the plasma periphery, and a steady component due to external coils. The method of choice is to solve for the linear response of the plasma (first order in the wave field amplitude) then evolve the time-averaged colliding distributions, viewed as slowly varying functions of time, according to the resulting time-averaged cooperative effects: The so-called "quasilinear" effects are second order in the wave amplitude.

Denote the operation of time-averaging a given dependent variable Q on the wave period by $\langle Q \rangle_t = \hat{Q}$, so that $Q = \hat{Q} + \tilde{Q}$; applying this operation to (3.1.1) there results the coupled pair of equations

$$\frac{\partial \hat{f}}{\partial t} + v \cdot \nabla \hat{f} + \nabla_v \cdot \left(\frac{q}{m} \left(\hat{E} + \frac{v}{c} \times \hat{B} \right) \hat{f} + \langle \Gamma_c \rangle_t + \langle \Gamma_{ql} \rangle_t \right) = 0 \qquad (3.1.4)$$

and

$$\frac{\partial \tilde{f}}{\partial t} + v \cdot \nabla \tilde{f} + \nabla_v \cdot \left(\frac{q}{m} \left(\hat{E} + \frac{v}{c} \times \hat{B} \right) \tilde{f} \right) = -\nabla_v \cdot \left(\frac{q}{m} \left(\tilde{E} + \frac{v}{c} \times \tilde{B} \right) \hat{f} \right). \qquad (3.1.5)$$

The quasilinear flux

$$\Gamma_{ql} = \frac{q}{m} \left(\tilde{E} + \frac{v}{c} \times \tilde{B} \right) \tilde{f}, \qquad (3.1.6)$$

which appears in (3.1.4), is to be determined through the solution of (3.1.5), $\tilde{f}$, and a knowledge of the fluctuating fields.

Suppose for the moment that the fluxes, time-averaged over a wave-period, are known. We are interested in the long-term time-averaged behavior of the distribution functions. Rapid fluctuations on intrinsic time scales such as wave-, gyro-, and bounce-periods are viewed as subordinate in this respect. We therefore undertake to perform the appropriate time-averages.

Since, unlike the wave-frequency, the gyro-frequency varies from point to point, and bounce-frequency varies from orbit to orbit, some care must be taken in the averaging procedure to preserve relevant detail. That is to say, we require the motion of the entity represented by the averaged equations to coincide with the average of the solutions to the exact equations; the latter being taken over an ensemble of initial conditions differing only with respect to the initial phase of the nearly recurrent motion with repect to which the averaging is to apply. The general procedure is to average first over the highest frequency nearly recurrent motion, then proceed to lower frequency orbit structures until the relevant (collisional) time scale is reached.

We presume that, in order of magnitude,

$$\nabla_{\mathbf{v}} \cdot \mathbf{\Gamma}_c \sim v_c f, \qquad \nabla_{\mathbf{v}} \cdot \mathbf{\Gamma}_{ql} \sim v_{ql} f, \tag{3.1.7}$$

and

$$\Omega \gg \omega_B \gg v_c \sim v_{ql}, \tag{3.1.8}$$

where Ω is the gyro-frequency, ω_B is the bounce-frequency, and v_c, v_{ql} are the corresponding interaction frequencies for particle–particle and wave–particle interaction.

The time-averaged Fokker–Planck equation (3.1.4) can be rewritten as

$$\frac{\partial \hat{f}}{\partial t} + \mathbf{v} \cdot \nabla \hat{f} + \dot{v} \frac{\partial \hat{f}}{\partial v} + \dot{\theta} \frac{\partial \hat{f}}{\partial \theta} + \dot{\phi} \frac{\partial \hat{f}}{\partial \phi} + \nabla_{\mathbf{v}} \cdot \langle \mathbf{\Gamma} \rangle_t = 0, \tag{3.1.9}$$

where the gyro-phase varies in time along a particle trajectory as

$$\dot{\phi} = \Omega + O(\Omega^0) + O(\Omega^{-1}) + \cdots \tag{3.1.10}$$

and the speed v and pitch-angle θ vary slowly as

$$\dot{v} \sim \dot{\theta} = O(\Omega^0) + O(\Omega^{-1}) + \cdots. \tag{3.1.11}$$

In large gyration frequency theory we seek a solution $\hat{f}$ of (3.1.9) in the form of a series ordered in inverse powers of the large frequency Ω;

$$\hat{f} = f + f_1 + f_2 + \cdots. \tag{3.1.12}$$

Substituting (3.1.12) into (3.1.9) and collecting terms at each order we find at lowest order that

$$\Omega \frac{\partial f}{\partial \phi} = 0. \tag{3.1.13}$$

The lowest order contribution is gyro-phase independent. At first order, (3.1.13) implies

$$-\Omega \frac{\partial f_1}{\partial \phi} = \frac{\partial f}{\partial t} + \mathbf{v} \cdot \nabla f + \dot{v} \frac{\partial f}{\partial v} + \dot{\theta} \frac{\partial f}{\partial \theta} + \nabla_{\mathbf{v}} \cdot \hat{\mathbf{\Gamma}}, \tag{3.1.14}$$

where $\hat{\mathbf{\Gamma}}$ is $\langle \mathbf{\Gamma} \rangle_t$ evaluated with $f = \hat{f}$. The function f_1 must be periodic in ϕ with period 2π. Define the gyro-phase average operation as

$$\langle Q \rangle_\phi = \frac{1}{2\pi} \int_0^{2\pi} Q \, d\phi. \tag{3.1.15}$$

Whence, upon gyro-phase averaging (3.1.14), there remains

$$0 = \left\langle \frac{\partial f}{\partial t} \right\rangle_\phi + \langle \mathbf{v} \cdot \nabla f \rangle_\phi + \left\langle \dot{v} \frac{\partial f}{\partial v} \right\rangle_\phi + \left\langle \dot{\theta} \frac{\partial f}{\partial \theta} \right\rangle_\phi + \langle \nabla_{\mathbf{v}} \cdot \hat{\mathbf{\Gamma}} \rangle_\phi. \tag{3.1.16}$$

Reduction of (3.1.16) proceeds as follows: Since f is gyro-phase independent,

$$\left\langle \frac{\partial f}{\partial t} \right\rangle_\phi = \frac{\partial f}{\partial t}. \tag{3.1.17}$$

The time on the right in this expression is now measured on a time scale slow with respect to a gyro-period, i.e., many gyro-periods in duration.

We introduce the co-moving velocity basis defined by

$$\hat{\mathbf{e}}_v = +\hat{\mathbf{b}} \cos \theta + \hat{\mathbf{e}}_1 \sin \theta \cos \phi + \hat{\mathbf{e}}_2 \sin \theta \sin \phi,$$
$$\hat{\mathbf{e}}_\theta = -\hat{\mathbf{b}} \sin \theta + \hat{\mathbf{e}}_1 \cos \theta \cos \phi + \hat{\mathbf{e}}_2 \cos \theta \sin \phi, \tag{3.1.18}$$

where $(\hat{\mathbf{b}}, \hat{\mathbf{e}}_1, \hat{\mathbf{e}}_2)$ is an orthonormal basis triad co-moving with the gyro-center and

$$\hat{\mathbf{b}} = \frac{\hat{\mathbf{B}}}{|\hat{\mathbf{B}}|}. \tag{3.1.19}$$

Higher order cross field drifts have been neglected in these guiding center co-moving coordinates. Using (3.1.18) it follows that

$$\langle \mathbf{v} \cdot \nabla f \rangle_\phi = v \langle \hat{\mathbf{e}}_v \rangle_\phi \cdot \nabla f = v \cos \theta \hat{\mathbf{b}} \cdot \nabla f. \tag{3.1.20}$$

Using $v = \hat{\mathbf{e}}_v \cdot \mathbf{v}$, we can evaluate $\langle \dot{v} \rangle_\phi$ as

$$\langle \dot{v} \rangle_\phi = \langle \dot{\mathbf{v}} \cdot \hat{\mathbf{e}}_v \rangle_\phi + \langle \mathbf{v} \cdot \dot{\hat{\mathbf{e}}}_v \rangle_\phi$$
$$= \left\langle \frac{q}{m} \left(\hat{\mathbf{E}} + \frac{\mathbf{v}}{c} \times \hat{\mathbf{B}} \right) \cdot \hat{\mathbf{e}}_v \right\rangle_\phi = \frac{q}{m} \hat{\mathbf{E}} \cdot \hat{\mathbf{b}} \cos \theta. \tag{3.1.21}$$

The pitch-angle coordinate is defined as

$$\theta = \arctan \left(\frac{v_\perp}{v_\parallel} \right), \tag{3.1.22}$$

where $v_\perp = |\mathbf{v} \times \hat{\mathbf{b}}| = v \sin \theta$ and $v_\parallel = \mathbf{v} \cdot \hat{\mathbf{b}} = v \cos \theta$. Therefore $\langle \dot{\theta} \rangle_\phi$ is given in these terms as

$$\langle \dot{\theta} \rangle_\phi = \left\langle \left[\arctan \left(\frac{v_\perp}{v_\parallel} \right) \right]^{\cdot} \right\rangle_\phi = \left\langle \frac{v_\parallel \dot{v}_\perp - v_\perp \dot{v}_\parallel}{v^2} \right\rangle_\phi. \tag{3.1.23}$$

But $v_\parallel$ and $v_\perp$ are constrained by the energy constraint

$$v_\parallel^2 + v_\perp^2 = v^2 \tag{3.1.24}$$

so that

$$v_\parallel \dot{v}_\parallel + v_\perp \dot{v}_\perp = v \dot{v}. \tag{3.1.25}$$

Using (3.1.25) to eliminate $\dot{v}_\perp$ in (3.1.23), $\langle \dot{\theta} \rangle_\phi$ can be represented as

$$\langle \dot{\theta} \rangle_\phi = \left\langle -\frac{\dot{v}_\parallel}{v_\perp} + \frac{v_\parallel \dot{v}}{v v_\perp} \right\rangle_\phi = -\frac{1}{v_\perp} \langle \dot{v}_\parallel \rangle_\phi + \frac{v_\parallel}{v v_\perp} \langle \dot{v} \rangle_\phi. \tag{3.1.26}$$

We have previously found $\langle \dot{v} \rangle_\phi$, (3.1.21), so it remains only to determine $v_\parallel$. By direct differentiation find

$$\langle \dot{v}_\parallel \rangle_\phi = \langle (\hat{\mathbf{b}} \cdot \mathbf{v})^{\cdot} \rangle_\phi = \langle \dot{\hat{\mathbf{b}}} \cdot \mathbf{v} \rangle_\phi + \langle \hat{\mathbf{b}} \cdot \dot{\mathbf{v}} \rangle_\phi. \tag{3.1.27}$$

A short calculation shows this evaluates to

$$\langle \dot{v}_\parallel \rangle_\phi = \left\langle \tfrac{1}{2} v_\perp^2 \nabla \cdot \hat{\mathbf{b}} + \mathbf{v}_\perp \cdot \left(\frac{\partial \hat{\mathbf{b}}}{\partial t} + v_\parallel \hat{\mathbf{b}} \cdot \nabla \hat{\mathbf{b}} \right) + v_\perp^2 \frac{\partial \mathbf{Q}}{\partial \phi} : \nabla \hat{\mathbf{b}} + \frac{q}{m} \hat{\mathbf{E}} \cdot \hat{\mathbf{b}} \right\rangle_\phi, \tag{3.1.28}$$

where

$$\mathbf{Q} = \tfrac{1}{4}(\hat{\mathbf{e}}_1 \hat{\mathbf{e}}_1 - \hat{\mathbf{e}}_2 \hat{\mathbf{e}}_2) \sin 2\phi - \tfrac{1}{4}(\hat{\mathbf{e}}_1 \hat{\mathbf{e}}_2 + \hat{\mathbf{e}}_2 \hat{\mathbf{e}}_1) \cos 2\phi, \tag{3.1.29}$$

and, bearing in mind the comment following (3.1.19).

$$\mathbf{v}_\perp = v_\perp (\hat{\mathbf{e}}_1 \cos \phi + \hat{\mathbf{e}}_2 \sin \phi). \tag{3.1.30}$$

Clearly, the only terms surviving the gyro-phase average of $\dot{v}_\parallel$ are

$$\langle \dot{v}_\parallel \rangle_\phi = \tfrac{1}{2} v_\perp^2 \nabla \cdot \hat{\mathbf{b}} + \frac{q}{m} \hat{\mathbf{E}} \cdot \hat{\mathbf{b}}. \tag{3.1.31}$$

Using (3.1.31) in (3.1.26) we have finally that

$$\langle \dot{\theta} \rangle_\phi = -\tfrac{1}{2} v_\perp \nabla \cdot \hat{\mathbf{b}} - \frac{q}{m} \hat{\mathbf{E}} \cdot \hat{\mathbf{b}} \frac{v_\perp}{v^2}. \tag{3.1.32}$$

Inserting (3.1.17), (3.1.20), (3.1.21), (3.1.32) in (3.1.16) there results the gyro-phase averaged gyro-kinetic equation

$$\frac{\partial f}{\partial t} + v \cos \theta \hat{\mathbf{b}} \cdot \nabla f - \frac{q}{m} \hat{\mathbf{E}} \cdot \hat{\mathbf{b}} \frac{\partial f}{\partial v_\parallel} - \tfrac{1}{2} v \sin \theta \nabla \cdot \hat{\mathbf{b}} \frac{\partial f}{\partial \theta} + \langle \nabla_\mathbf{v} \cdot \hat{\Gamma} \rangle_\phi = 0. \tag{3.1.33}$$

Under most circumstances, field variation is sufficiently weak to assure that the changes in the parameters characterizing the unperturbed motion are adiabatic to requisite order. As a consequence there may exist nearly recurrent motion on an extended time scale (many gyro-periods) and certain invariants associated with motion along the field as well as the usual invariants associated with the gyro-motion. In a fashion analogous to the foregoing, we presume the presence of a magnetic well satisfying these conditions and nearly recurrent motion associated with this well of frequency ω_B greater than any other characteristic frequency of interest except the gyro-frequency. This bounce-frequency, ω_B, with which a gyro-center executes its nearly recurrent orbital motion in the varying field structure, is given by the path integral

$$\frac{2\pi}{\omega_B} = \tau_B = \oint \frac{ds}{v \cos \theta} = \oint \frac{ds}{v_\parallel}, \tag{3.1.34}$$

where ds is the element of arclength along the magnetic field line associated with the gyro-center motion. The relevant bounce-phase infinitesmal $d\phi_B$ is defined by

$$d\phi_B = \omega_B \frac{ds}{v \cos \theta}, \tag{3.1.35}$$

so that

$$\oint d\phi_B = 2\pi. \tag{3.1.36}$$

Recognizing that $\hat{\mathbf{b}} \cdot \nabla$ is the operator of differentiation with respect to arclength along field lines, and using (3.1.35), we can make the identification

$$v \cos \theta \hat{\mathbf{b}} \cdot \nabla = v \cos \theta \frac{d}{ds} = \omega_B \frac{\partial}{\partial \phi_B}. \tag{3.1.37}$$

Futhermore, since $\nabla \cdot \mathbf{B} = 0$, we can recast the fourth term in (3.1.33) as

$$-\tfrac{1}{2} v \sin \theta \nabla \cdot \hat{\mathbf{b}} \frac{\partial f}{\partial \theta} = \frac{\mu}{mv \sin \theta} \frac{d\hat{B}}{ds} \frac{\partial f}{\partial \theta}, \tag{3.1.38}$$

where the magnetic moment

$$\mu = \frac{mv^2 \sin^2 \theta}{2B} \tag{3.1.39}$$

is an invariant of the unperturbed motion. Defining

$$\psi = \frac{B(s)}{B(s = 0)}, \tag{3.1.40}$$

the invariance of μ can be restated as

$$\sin^2 \theta = \psi \sin^2 \theta_0, \tag{3.1.41}$$

where θ_0 is the pitch-angle coordinate at a fixed point ($s = 0$) in the field through which all orbits pass. We adopt the notation that subscript $_0$ hereafter will refer to quantities evaluated at this point which we call the midplane. In addition, we specialize to field structures which are periodic and piecewise monotonic about the minimum point at $s = 0$, continuous and continuously differentiable everywhere. This restriction is sufficiently lax to include the major features of a wide variety of magnetic fusion devices in which there exists a magnetic well. Collisions, interaction with microwave fields, and ohmic electric fields are to be viewed as perturbative[1] in this context so that to requisite order

---

[1] Perturbative: the energy gained or lost by a particle as a result of motion along the parallel electric field is viewed as small, i.e., of the order of the energy gained or lost in collisions with particles and interaction with waves.

$$v = v_0, \tag{3.1.42}$$

the speed is also a constant of the unperturbed motion. Upon application of the chain rule for differentiation, and using (3.1.41) and (3.1.42), we transform the pitch-angle derivative as

$$\frac{\partial}{\partial \theta} = \frac{\partial \theta_0}{\partial \theta} \frac{\partial}{\partial \theta_0} + \frac{\partial v_0}{\partial \theta} \frac{\partial}{\partial v_0} = \frac{\cos \theta}{\sqrt{\psi} \cos \theta_0} \frac{\partial}{\partial \theta_0}, \tag{3.1.43}$$

whereupon with (3.1.37), (3.1.38) we can recast (3.1.33) in the midplane variables v_0, θ_0 in the form

$$\frac{\partial f}{\partial t} + \omega_B \frac{\partial f}{\partial \phi_B} + \frac{\omega_B B_0 \mu \partial \ln \psi / \partial \phi_B}{m v^2 \sin \theta_0 \cos \theta_0} \frac{\partial f}{\partial \theta_0} = -\langle \nabla_v \cdot \hat{\Gamma} \rangle_\phi - \frac{q}{m} \hat{E} \cdot \hat{b} \frac{\partial f}{\partial v_\parallel}, \tag{3.1.44}$$

Proceeding along similar lines as in large gyration frequency theory (cf. (3.1.12)), presuming an expansion for f in inverse powers of the large bounce-frequency, ω_B,

$$f = \mathscr{F} + \mathscr{F}_1 + \mathscr{F}_2 + \cdots, \tag{3.1.45}$$

the bounce-average of (3.1.44) determines that

$$\omega_B \frac{\partial \mathscr{F}}{\partial \phi_B} = 0, \tag{3.1.46}$$

whence $\mathscr{F}$ is bounce-phase independent. By independent here we mean there is no explicit dependence on ϕ_B, the implicit dependence arising through the orbit equations. The equations of motion of the particle guiding centers provide the transformation relating the distribution function to be evolved at a given s or ϕ_B with its counterpart at any other s as follows:

$$\mathscr{F}_0(v_0, \theta_0) = \mathscr{F}(v(v_0, \theta_0; s), \theta(v_0, \theta_0; s)), \tag{3.1.47}$$

where we have previously defined v_0, θ_0 to be the speed and pitch-angle in the equatorial midplane of a particle whose speed and pitch-angle are v, θ at arclength s along the field. At next order

$$-\omega_B \frac{\partial \mathscr{F}_1}{\partial \phi_B} = \frac{\partial \mathscr{F}}{\partial t} + \frac{\omega_B B_0 \mu \partial \ln \psi / \partial \phi_B}{m v^2 \sin \theta_0 \cos \theta_0} \frac{\partial \mathscr{F}}{\partial \theta_0} + \langle \nabla_v \cdot \hat{\Gamma} \rangle_\phi + \frac{q}{m} \hat{E} \cdot \hat{b} \frac{\partial \mathscr{F}}{\partial v_\parallel}. \tag{3.1.48}$$

Now again, since $\mathscr{F}_1$ is periodic in ϕ_B of period 2π, upon averaging over ϕ_B there remains the evolution equation for $\mathscr{F}$ where now t is measured on a time scale slow with respect to a bounce-period, viz. many bounce-periods in duration:

$$\frac{\partial \mathscr{F}}{\partial t} = -\langle\langle \nabla_v \cdot \hat{\Gamma} \rangle_\phi \rangle_{\phi_B} - \left\langle \frac{q}{m} \hat{E} \cdot \hat{b} \frac{\partial \mathscr{F}}{\partial v_\parallel} \right\rangle_{\phi_B}. \tag{3.1.49}$$

Henceforth, for ease of notation, we will denote the double phase average indicated in (3.1.49) simply by enclosing the expression to be averaged in $\langle\!\langle \, \rangle\!\rangle$.

Let us now consider the right-hand side of this bounce-averaged Fokker–Planck equation. The gyro-phase averaged Fokker–Planck operator in a uniform magnetic field has been treated extensively. In the inhomogeneous magnetic field, the analogous local (fixed s) Fokker–Planck operator can be written in the same form

$$
\begin{aligned}
\left(\frac{\partial \mathscr{F}}{\partial t}\right)_{cql} &= -\langle \nabla_v \cdot \hat{\Gamma}\rangle_\phi \\
&= \gamma\left(\frac{1}{v^2}\frac{\partial}{\partial v}\left(A + B\frac{\partial}{\partial v} + C\frac{\partial}{\partial\theta}\right) + \frac{1}{v^2\sin\theta}\frac{\partial}{\partial\theta}\left(D + E\frac{\partial}{\partial v} + F\frac{\partial}{\partial\theta}\right)\right)\mathscr{F} \\
&= \gamma\left(\frac{1}{v^2}\frac{\partial\mathscr{G}}{\partial v} + \frac{1}{v^2\sin\theta}\frac{\partial\mathscr{H}}{\partial\theta}\right),
\end{aligned}
\tag{3.1.50}
$$

where the flux $\hat{\Gamma}$ is composed of the three parts

$$
\hat{\Gamma} = \hat{\Gamma}_c + \hat{\Gamma}_{ql} + \hat{\Gamma}_e
\tag{3.1.51}
$$

due, respectively, to the effects of Coulomb collisions, wave-particle interaction, and ohmic electric field. The operator (3.1.50) is to be viewed as representing the time rate of change of the local distribution of particles in velocity space due to local interactions at orbit point s (or equivalently, to requisite order, at bounce-phase ϕ_B). Equation (3.1.50) is not, however, an evolution equation for the bounce-averaged distribution function; it is merely a representation of the operator which must be averaged according to (3.1.15) and then used in (3.1.49).

Regarding (3.1.49) and the comments thereafter, we define the notation

$$
\begin{aligned}
\left\langle\!\!\left\langle \left(\frac{\delta\mathscr{F}_0}{\partial t}\right)_{cql}\right\rangle\!\!\right\rangle &= \frac{1}{\tau_B}\int_0^{s_B}\frac{ds}{v_\parallel}\left(\frac{\delta\mathscr{F}_0}{\delta t}\right)_{cql} = \frac{1}{2\pi}\int_0^{2\pi}\left(\frac{\delta\mathscr{F}_0}{\delta t}\right)_{cql}d\phi_B \\
&= \langle\!\langle \nabla_v \cdot \hat{\Gamma}\rangle\!\rangle = \frac{1}{v_0\cos\theta_0\tau_B}\nabla_{v_0}\cdot\hat{\Gamma}_0.
\end{aligned}
\tag{3.1.52}
$$

Here we take the bounce-average over one-quarter of a cycle of the bounce-motion and the analogous motion for transiting particles so that s_B in a tokamak is given by

$$
\begin{aligned}
\cos(\theta(v_0, \theta_0; s_B)) &= \cos(\theta_T) = 0; & \text{for trapped particles,} \\
s_B = s_{mx} &= \pi q R_0; \quad q = rB_T/R_0B_\theta; & \text{for circulating particles,}
\end{aligned}
\tag{3.1.53}
$$

and where q is the safety factor.

The operator (3.1.52) can be decomposed as was the flux $\hat{\Gamma}$ in (3.1.50); each part conserves (line density) N separately. That is to say, it can be shown (see Appendix 3A) that

$$0 = \frac{\partial N}{\partial t} = \int d^3 v_0 \, v_0 \cos \theta_0 \tau_B \left(\frac{\delta \mathcal{F}}{\delta t} \right)_{cql} = \int d^3 v_0 \, v_0 \cos \theta_0 \tau_B \langle\!\langle \nabla_v \cdot \hat{\Gamma} \rangle\!\rangle$$

$$= \int d^3 v_0 \, \nabla_{v_0} \cdot \hat{\Gamma}_0,$$

(3.1.54)

where ∇_{v_0} is the gradient operator in the midplane velocity coordinate system. $\hat{\Gamma}_0$ is found by rewriting (3.1.50) as

$$\left(\frac{\delta \lambda \mathcal{F}_0}{\delta t} \right)_{cql} = -\nabla_{v_0} \cdot \hat{\Gamma}_0$$

$$= \gamma \left(\frac{1}{v_0^2} \frac{\partial}{\partial v_0} \left(A_0 + B_0 \frac{\partial}{\partial v_0} + C_0 \frac{\partial}{\partial \theta_0} \right) \right.$$

$$\left. + \frac{1}{v_0^2 \sin \theta_0} \frac{\partial}{\partial \theta_0} \left(D_0 + E_0 \frac{\partial}{\partial v_0} + F_0 \frac{\partial}{\partial \theta_0} \right) \right) \mathcal{F}_0$$

(3.1.55)

$$= \gamma \left(\frac{1}{v_0^2} \frac{\partial \mathcal{G}_0}{\partial v_0} + \frac{1}{v_0^2 \sin \theta_0} \frac{\partial \mathcal{H}_0}{\partial \theta_0} \right),$$

where the coefficients $A_0 \to F_0$ can be found with prior knowledge of the coefficients $A \to F$ (3.1.50), and $\lambda = v_0 \cos \theta_0 \tau_B$.

Now since the particle transit or bounce time is small compared with a collisional relaxation time we evaluate changes in the bounce-averaged distribution $\mathcal{F}$ through an integration of its time rate of change along orbits in phase space. The procedure followed to obtain the coefficients $A_0 \to F_0$ in (3.1.55) thus involves two steps:

(A) Determine the Fokker–Planck coefficients at various points in s along a guiding-center orbit in a manner analogous to that given in the conventional uniform field treatment.
(B) Bounce-average these coefficients by accumulating a weighted sum (i.e., perform the integral (3.1.52)) suitable for time advancing the distribution $\mathcal{F}_0$.

From (3.1.50) we write the local collision operator for species type "a" at orbit point s as

$$\left(\frac{\delta \mathcal{F}_a}{\delta t} \right)_c = \gamma_a \left(\frac{1}{v^2} \frac{\partial \mathcal{G}_a}{\partial v} + \frac{1}{v^2 \sin \theta} \frac{\partial \mathcal{H}_a}{\partial \theta} \right),$$

(3.1.56)

where we have reintroduced the speed and pitch-angle velocity space collisional fluxes $\mathcal{G}_a$ and $\mathcal{H}_a$, as in (3.1.50), given by

$$\mathcal{G}_a = A_a \mathcal{F}_a + B_a \frac{\partial \mathcal{F}_a}{\partial v} + C_a \frac{\partial \mathcal{F}_a}{\partial \theta},$$

(3.1.57)

and

$$\mathcal{H}_a = D_a\mathcal{F}_a + E_a\frac{\partial \mathcal{F}_a}{\partial v} + F_a\frac{\partial \mathcal{F}_a}{\partial \theta}. \tag{3.1.58}$$

Note that in the above a sum over species is implied, and $\gamma_a = 4\pi Z_a^4 e^4/m_a^2$.

These equations are identical in form to their analogs in uniform field theory, it being understood that $\mathcal{F}$ here is the bounce-phase independent component of f, the gyro-phase averaged distribution, at the orbit point parametrized by arclength s, and velocity space coordinates $(v(v_0, \theta_0, s), \theta(v_0, \theta_0, s))$ (the subscript is the species identifier). Note that although $\mathcal{F}$ and therefore also $\mathcal{G}$ and $\mathcal{H}$ are implicitly dependent upon parameter s, they are nevertheless bounce-phase independent, viz. (3.1.47) applies.

The coefficients $A_a \to F_a$ are given by equations again identical in form to their analogs in uniform field theory:

$$A_a = \frac{v^2}{2}\frac{\partial^3 g_a}{\partial v^3} + v\frac{\partial^2 g_a}{\partial v^2} - \frac{\partial g_a}{\partial v} - v^2\frac{\partial h_a}{\partial v} - \frac{1}{v}\frac{\partial^2 g_a}{\partial \theta^2}$$

$$+ \frac{1}{2}\frac{\partial^3 g_a}{\partial v \partial \theta^2} - \frac{\cot\theta}{v}\frac{\partial g_a}{\partial \theta} + \frac{\cot\theta}{2}\frac{\partial^2 g_a}{\partial v \partial \theta},$$

$$B_a = \frac{v^2}{2}\frac{\partial^2 g_a}{\partial v^2},$$

$$C_a = -\frac{1}{2v}\frac{\partial g_a}{\partial \theta} + \frac{1}{2}\frac{\partial^2 g_a}{\partial v \partial \theta},$$

$$D_a = \frac{\sin\theta}{2v^2}\frac{\partial^3 g_a}{\partial \theta^3} + \frac{\sin\theta}{2}\frac{\partial^3 g_a}{\partial v^2 \partial \theta} + \frac{\sin\theta}{v}\frac{\partial^2 g_a}{\partial v \partial \theta} - \frac{1}{2v^2 \sin\theta}\frac{\partial g_a}{\partial \theta}$$

$$+ \frac{\cos\theta}{2v^2}\frac{\partial^2 g_a}{\partial \theta^2} - \sin\theta\frac{\partial h_a}{\partial \theta},$$

$$E_a = -\frac{\sin\theta}{2v}\frac{\partial g_a}{\partial \theta} + \frac{\sin\theta}{2}\frac{\partial^2 g_a}{\partial v \partial \theta},$$

$$F_a = \frac{\sin\theta}{2v^2}\frac{\partial^2 g_a}{\partial \theta^2} + \frac{\sin\theta}{2v}\frac{\partial g_a}{\partial v}. \tag{3.1.59}$$

The mathematical procedure employed in the generation of the local (in s) Fokker–Planck coefficients (3.1.59) is similar to that described in Chapter 2 with suitable consideration given the fact there is an additional parameter (s).

The functions g_a and h_a are the local multispecies Rosenbluth potentials given by

$$g_a = \sum_a \frac{Z_b^2}{Z_a^2} \ln \Lambda_{ab} \int d\mathbf{v}' \, \mathcal{F}_b(\mathbf{v}')|\mathbf{v} - \mathbf{v}'|, \tag{3.1.60}$$

and

$$h_a = \sum_b \frac{m_a + m_b}{m_b} \frac{Z_b^2}{Z_a^2} \ln \Lambda_{ab} \int d\mathbf{v}' \, \mathscr{F}_b(\mathbf{v}')|\mathbf{v} - \mathbf{v}'|^{-1}, \qquad (3.1.61)$$

where $\ln \Lambda_{ab}$ is the Coulomb logarithm. The integrals are over the entire domain in velocity space at point s. The local "Rosenbluth potentials" and the distribution functions themselves may be represented by expansions in Legendre polynomials. For this purpose we let

$$\mathscr{F}_b(v, \theta, s; t) = \sum_{m=0}^{\infty} V_m^b(v, s; t) P_m(\cos \theta), \qquad (3.1.62)$$

where the Fourier–Legendre coefficients of the local distribution function are given by the projection integrals

$$V_m^b(v, s; t) = \frac{2m + 1}{2} \int_{-1}^{+1} d(\cos \theta) \mathscr{F}_b(v, \cos \theta, s; t) P_m(\cos \theta). \qquad (3.1.63)$$

In practice, this integral is evaluated by transforming the variable of integration to the midplane system and using the appropriately transformed limits. The restricted domain of integration in the midplane system is that portion of (v_0, θ_0) space which maps into $(v(v_0, \theta_0, s), \theta(v_0, \theta_0, s))$ space at the relevant point s. Particles which bounce before reaching point s make no contribution to the collisions occurring there. The expansions for the potentials can be written

$$g_a(v, \theta, s; t) = \sum_{m=0}^{\infty} \sum_b \frac{Z_b^2}{Z_a^2} \ln \Lambda_{ab} B_m^b(v, s; t) P_m(\cos \theta), \qquad (3.1.64)$$

and

$$h_a(v, \theta, s; t) = \sum_{m=0}^{\infty} \sum_b \frac{m_a + m_b}{m_b} \frac{Z_b^2}{Z_a^2} \ln \Lambda_{ab} A_m^b(v, s; t) P_m(\cos \theta), \qquad (3.1.65)$$

where the local Rosenbluth potential Fourier–Legendre coefficients are related to the distribution function Fourier–Legendre coefficients through

$$A_m^b = \frac{4\pi}{2m + 1} \left(\int_0^v dv' \frac{v'^{m+2}}{v^{m+1}} V_m^b(v', s; t) + \int_v^{\infty} dv' \frac{v^m}{v'^{m-1}} V_m^b(v', s; t) \right), \qquad (3.1.66)$$

and

$$B_m^b = -\frac{4\pi}{4m^2 - 1} \left\{ \int_0^v dv' \frac{v'^{m+2}}{v^{m-1}} \left(1 - \frac{m - \frac{1}{2} v'^2}{m + \frac{3}{2} v^2} \right) V_m^b(v', s; t) \right.$$
$$\left. + \int_v^{\infty} dv' \frac{v^m}{v'^{m-3}} \left(1 - \frac{m - \frac{1}{2} v^2}{m + \frac{3}{2} v'^2} \right) V_m^b(v', s; t) \right\}. \qquad (3.1.67)$$

Again, as in uniform field theory, it is notationally convenient to use the four functionals M_m, N_m, R_m, and E_m defined in (2.1.25)–(2.1.28)

$$M_m(y; v) = \int_v^{\infty} y(\xi) \xi^{1-m} \, d\xi,$$

$$N_m(y; v) = \int_0^v y(\xi)\xi^{2+m}\, d\xi,$$

$$R_m(y; v) = \int_v^\infty y(\xi)\xi^{3-m}\, d\xi,$$

$$E_m(y; v) = \int_0^v y(\xi)\xi^{4+m}\, d\xi. \tag{3.1.68}$$

Noting that these functionals now depend implicitly on s through the magnetic field inhomogeneity not present in the previous calculation of Chapter 2, (3.1.66) and (3.1.67) become

$$A_m^b = \frac{4\pi}{2m + 1}(v^{-m-1}N_m(V_m^b) + v^m M_m(V_m^b)), \tag{3.1.69}$$

and

$$B_m^b = \frac{4\pi}{2m + 1}\left\{ \frac{1}{2m + 3}(v^{-m-1}E_m(V_m^b) + v^{m+2}M_m(V_m^b)) \right.$$

$$\left. - \frac{1}{2m - 1}(v^{1-m}N_m(V_m^b) + v^m R_m(V_m^b)) \right\}. \tag{3.1.70}$$

Derivatives of the A_m^b and B_m^b with respect to v, taken at constant s, are also expressible in terms of the same four functionals as given by (2.1.31)–(2.1.34). Derivatives with respect to θ are obtained analytically through differentiation of the Legendre polynomials.

The coefficients $A_a(v, \theta, s; t) \rightarrow F_a(v, \theta, s; t)$ are computed at values of θ for which there exists a $\theta_0(i)$ (on the midplane mesh) such that $\theta = \theta(\theta_0(i), s)$: Coefficients are determined along orbits piercing the midplane at meshpoints; consequently, when the bounce-average is performed no interpolation is required. As mentioned previously, the quadratures (e.g., (3.1.63)) are transformed to the midplane, with the help of (3.1.47): The limits of integration are determined parametrically by s.

Using the notation introduced in (3.1.52) to denote the bounce-averaging operation and the definition of λ following (3.1.55), as is shown in Appendix 3A, we can represent the coefficients $A_0 \rightarrow F_0$ in the bounce-averaged collision operator (3.1.55) in conservation form as

$$A_0 = \lambda \langle\!\langle A \rangle\!\rangle,$$

$$B_0 = \lambda \langle\!\langle B \rangle\!\rangle,$$

$$C_0 = \lambda \left\langle\!\!\left\langle \frac{\cos\theta}{\sqrt{\psi}\,\cos\theta_0} C \right\rangle\!\!\right\rangle,$$

$$D_0 = \lambda \left\langle\!\!\left\langle \frac{\cos\theta}{\psi\,\cos\theta_0} D \right\rangle\!\!\right\rangle,$$

$$E_0 = \lambda \left\langle\!\!\left\langle \frac{\cos\theta}{\psi\cos\theta_0} E \right\rangle\!\!\right\rangle,$$

$$F_0 = \lambda \left\langle\!\!\left\langle \frac{\cos^2\theta}{(\sqrt{\psi})^3\cos^2\theta_0} F \right\rangle\!\!\right\rangle. \tag{3.1.71}$$

In the event that the effects of a DC parallel electric field are to be included, the coefficients A_0 and D_0 must be augmented for passing orbits (see Section 3.1.4) by

$$A_{E0} = -s^* E_{\|0} \frac{q}{m} v_0^2 \cos\theta_0, \tag{3.1.72}$$

and

$$D_{E0} = s^* E_{\|0} \frac{q}{m} v_0 \sin^2\theta_0, \tag{3.1.73}$$

where $s^* = \int ds\,\psi$.

3.1.2. Bounce-averaged resonant diffusion

There is a growing body of literature dealing with the subject of quasilinear diffusion in a magnetized plasma. The treatment of Bernstein and Baxter (1981) is particularly instructive and the following analysis parallels its development to some extent.

In order to derive a representation for the quasilinear velocity current density, $\hat{\Gamma}_{ql}$, it is necessary to solve (3.1.5), as is apparent from (3.1.6). Ideally, the wave field is known self-consistently through a geometrical or physical optics calculation properly involving the particle trajectories in the reactive and dissipative parts of the wave dispersion properties of the plasma medium. For the moment, however, suppose the wave field to be known.

Then, to solve (3.1.5), presume a solution to the unperturbed orbit equations

$$\frac{d\mathbf{r}'}{d\tau} = \mathbf{v}', \tag{3.1.74}$$

and

$$\frac{d\mathbf{v}'}{d\tau} = \frac{q}{m}(\hat{\mathbf{E}}(\mathbf{r}', \tau) + \frac{\mathbf{v}'}{c} \times \hat{\mathbf{B}}(\mathbf{r}', \tau)) \tag{3.1.75}$$

subject to the initial conditions $\mathbf{r}'(\mathbf{r}, \mathbf{v}, t) = \mathbf{r}$ and $\mathbf{v}'(\mathbf{r}, \mathbf{v}, t) = \mathbf{v}$. Equation (3.1.5) can therefore be expressed as

$$\frac{d\tilde{f}}{d\tau} = \frac{\partial\tilde{f}}{\partial t} + \mathbf{v}\cdot\nabla\tilde{f} + \nabla_\mathbf{v}\cdot\left(\frac{q}{m}\left(\hat{\mathbf{E}} + \frac{\mathbf{v}}{c}\times\hat{\mathbf{B}}\right)\tilde{f}\right) = -\nabla_\mathbf{v}\cdot\left(\frac{q}{m}\left(\tilde{\mathbf{E}} + \frac{\mathbf{v}}{c}\times\tilde{\mathbf{B}}\right)\bar{f}\right). \tag{3.1.76}$$

Choose the ansatz $\tilde{\mathbf{E}}' = \tilde{\mathbf{E}}(\mathbf{r}', \tau) = \mathbf{h}(\mathbf{r}')\exp(-i\omega\tau)$ for the fluctuating RF electric field so that the fluctuating magnetic field is given by $\tilde{\mathbf{B}}' = -(ic/\omega)\nabla \times$

$\mathbf{h}\exp(-i\omega\tau)$. The solution to (3.1.76), the first-order fluctuating distribution function, can be represented as the time integral

$$\tilde{f}(\mathbf{r},\mathbf{v};t) = -\operatorname{Re}\int_{-\infty}^{t} d\tau \frac{q}{m}\left(\mathbf{h} - \frac{i}{\omega}\mathbf{v}\times\nabla\times\mathbf{h}\right)\cdot\nabla_{\mathbf{v}}\hat{f}e^{-i\omega\tau} = \operatorname{Re}(e^{-i\omega t}g).$$

(3.1.77)

Next let $\mathbf{r} = \mathbf{R} + \rho$ where $\mathbf{R}$ is the position of the guiding center and recast the equations of motion (3.1.74)–(3.1.75) as equations governing the motion of the guiding center:

$$\frac{d\mathbf{R}'}{d\tau} = v'_{\parallel}\hat{\mathbf{b}} + \frac{\hat{\mathbf{b}}'}{\Omega'}\times\left(\frac{\mu}{m}\nabla'B'_0 + v'^2_{\parallel}\hat{\mathbf{b}}'\cdot\nabla'\hat{\mathbf{b}}' - \frac{q}{m}\mathbf{E}'_0\right) = v'_{\parallel}\hat{\mathbf{b}}' + \mathbf{v}'_{\text{drift}}, \quad (3.1.78)$$

and

$$\frac{dv'_{\parallel}}{d\tau} = -\hat{\mathbf{b}}'\cdot\nabla'(\mu B'_0), \quad (3.1.79)$$

where $\hat{\mathbf{b}}'$, defined in (3.1.19), is a unit vector in the direction of the magnetic field (at $\mathbf{R}'$), $\mathbf{v}' = \mathbf{v}'\cdot\hat{\mathbf{b}}'\hat{\mathbf{b}}' + \hat{\mathbf{b}}'\times(\mathbf{v}'\times\hat{\mathbf{b}}') = v'_{\parallel}\hat{\mathbf{b}}' + \mathbf{v}'_{\perp}$ is the particle velocity, and $\Omega' = qB'/mc$ is the local gyro-frequency. The gyro-position vector of the particle is given by

$$\rho = \sqrt{\frac{2\mu|\mathbf{B}|}{\Omega^2 m}}\{\hat{\mathbf{e}}_1 \sin(\phi + \Phi(\tau) - \Phi(t)) + \mathbf{e}_2 \cos(\phi + \Phi(\tau) - \Phi(t))\}, \quad (3.1.80)$$

where $\hat{\mathbf{e}}'_1$ and $\hat{\mathbf{e}}'_2$ are mutually orthogonal unit vectors orthogonal to $\hat{\mathbf{b}}'$ at $\mathbf{R}'$ and

$$\Phi(t) = \int_0^t d\tau\ \Omega'(R'(\tau)). \quad (3.1.81)$$

Particle orbits differing only with respect to ϕ constitute a Kruskal ring (Kruskal, 1962): presently we shall perform the appropriate average eliminating this gyro-phase dependence.

Since we have in mind eventually to parallel this calculation with a geometrical optics calculation of the wave fields, we choose $\mathbf{h}(\mathbf{r})$ to have the form

$$\mathbf{h}(\mathbf{r}) = \mathbf{a}(\mathbf{r})\,e^{i\Lambda(\mathbf{r})}, \quad (3.1.82)$$

where $\mathbf{a}(\mathbf{r})$ is a slowly varying (in space) amplitude and the local wavevector is $\mathbf{k} = \nabla\Lambda$.

Along the particle orbit the eikonal Λ varies according to

$$\frac{d\Lambda}{d\tau} = \frac{d\Lambda}{d\mathbf{r}'}\cdot\frac{d\mathbf{r}'}{d\tau} = \frac{d\Lambda}{d\mathbf{r}'}\cdot\frac{d}{d\tau}(\mathbf{R}' + \rho'). \quad (3.1.83)$$

Using (3.1.78) and (3.1.80) in (3.1.83) and neglecting the contributions of terms which have only slow variation along the orbit, (3.1.83) integrates to

$$\Lambda(\mathbf{r}') = \Lambda(\mathbf{r}) + \mathbf{k}' \cdot \boldsymbol{\rho}' - \mathbf{k} \cdot \boldsymbol{\rho} + \int_t^\tau d\tau' \, (\hat{\mathbf{b}}' \cdot \mathbf{k}' v_\parallel' + \mathbf{k}' \cdot \mathbf{v}'_{\text{drift}}). \qquad (3.1.84)$$

Inserting this expression for the eikonal in (3.1.77), the first-order distribution function becomes

$$g(\mathbf{r}, \mathbf{v}; t) = -\text{Re} \sum_{\text{rays}} \int_{-\infty}^t d\tau \frac{q}{m} \left(\mathbf{a}' + \frac{\mathbf{v}'}{\omega} \times \mathbf{k}' \times \mathbf{a}' \right) \cdot \nabla_{\mathbf{v}'} \hat{f}$$

$$\times \exp\left(i\left(\Lambda(\mathbf{r}) + \int_t^\tau d\tau' \, (\hat{\mathbf{b}}' \cdot \mathbf{k}' v_\parallel' + \mathbf{k}' \cdot \mathbf{v}'_{\text{drift}} - \omega) + \mathbf{k}' \cdot \boldsymbol{\rho}' - \mathbf{k} \cdot \boldsymbol{\rho} \right) \right), \qquad (3.1.85)$$

where the summation $\sum_{\text{rays}}$ denotes a sum over all geometrical optics rays converging on the point in question.

Considering the guiding center drift velocity to be of higher order, the velocity space gradient appearing in (3.1.85) can be expanded as

$$\nabla_{\mathbf{v}} = \nabla_{\mathbf{v}} |\mathbf{v}| \frac{\partial}{\partial v} + \nabla_{\mathbf{v}} \theta \frac{\partial}{\partial \theta} = \hat{\mathbf{e}}_v \frac{\partial}{\partial v} + \hat{\mathbf{e}}_\theta \frac{1}{v} \frac{\partial}{\partial \theta}, \qquad (3.1.86)$$

where we can represent the velocity space unit vectors $\hat{\mathbf{e}}_v$ and $\hat{\mathbf{e}}_\theta$ in the basis $(\hat{\mathbf{b}}, \hat{\mathbf{e}}_1, \hat{\mathbf{e}}_2)$ following (3.1.18), as follows:

$$\hat{\mathbf{e}}_v = +\hat{\mathbf{b}} \cos\theta + \hat{\mathbf{e}}_1 \sin\theta \cos\alpha + \hat{\mathbf{e}}_2 \sin\theta \sin\alpha,$$

$$\hat{\mathbf{e}}_\theta = -\hat{\mathbf{b}} \sin\theta + \hat{\mathbf{e}}_1 \cos\theta \cos\alpha + \hat{\mathbf{e}}_2 \cos\theta \sin\alpha, \qquad (3.1.87)$$

where along the particle trajectory

$$\alpha = \Phi(\tau) - \Phi(t) + \phi. \qquad (3.1.88)$$

Representing the wavevector $\mathbf{k}$ in the local basis as

$$\mathbf{k} = k_\parallel \hat{\mathbf{b}} + k_\perp (\hat{\mathbf{e}}_1 \cos\delta + \hat{\mathbf{e}}_2 \sin\delta) \qquad (3.1.89)$$

and using (3.1.87)–(3.1.89) in (3.1.85), that expression for g is carried into

$$g = -\frac{q}{m} \sum_{\text{rays}} \sum_n e^{i(\Lambda - \mathbf{k} \cdot \boldsymbol{\rho} - \chi + n\phi)} \int_{-\infty}^t d\tau \, e^{i(\chi' - n\delta')}$$

$$\times \mathbf{a}' \cdot \left\{ \hat{\mathbf{b}}' \frac{\cos\theta'}{\sin\theta'} J_n' \left[\sin\theta' \frac{\partial}{\partial v'} + \left(\frac{\cos\theta'}{v'} - \frac{1 - n\Omega'/\omega}{v' \cos\theta'} \right) \frac{\partial}{\partial \theta'} \right] \right.$$

$$+ [\hat{\mathbf{e}}_+' e^{+i\delta'} J_{n-1}' + \hat{\mathbf{e}}_-' e^{-i\delta'} J_{n+1}']$$

$$\times \left. \left[\sin\theta' \frac{\partial}{\partial v'} + \left(\frac{\cos\theta'}{v'} - \frac{k_\parallel'}{\omega} \right) \frac{\partial}{\partial \theta'} \right] \right\} \hat{f}, \qquad (3.1.90)$$

where we have employed the identity

$$e^{iz \sin\alpha} = \sum_{n=-\infty}^\infty J_n(z) \, e^{in\alpha} \qquad (3.1.91)$$

and the argument of J_n' is $k_\perp' \rho'$. The summation in (3.1.91) is over all n, $-\infty < n < +\infty$, and we have introduced the symbol χ to signify

$$\chi(n, t) = \int_0^t d\tau \, (n\Omega' + k'_\parallel v'_\parallel - \omega) = \int_0^t d\tau \, v(n, \tau). \qquad (3.1.92)$$

Also introduced in (3.1.90) are the (complex) polarization vectors $\hat{\mathbf{e}}_\pm = (\hat{\mathbf{e}}_1 \pm i\hat{\mathbf{e}}_2)/2$. The time-averaged fluctuating distribution function g in the form (3.1.90) constitutes the solution of (3.1.5) sought for the purpose of representing the quasilinear velocity space flux (3.1.6) which appears in the quasilinear Fokker-Planck equation (3.1.4), whence, by a short calculation we find

$$\langle \hat{\Gamma}_{ql} \rangle_t = \frac{q}{4m} \sum_{\text{rays}} \left(\mathbf{a} + \frac{\mathbf{v}}{\omega} \times \mathbf{k} \times \mathbf{a} \right)^* e^{-i\Lambda} g + cc. \qquad (3.1.93)$$

The symbol * has been used to signify complex conjugation.

Replacing the expression (3.1.90) for g in (3.1.93) and expanding $\exp(i\mathbf{k} \cdot \rho)$ as in (3.1.91) we come to an expression of the time-averaged quasilinear velocity space flux which is directly amenable to gyro-phase averaging. Taking the $\hat{\mathbf{e}}_v$ and $\hat{\mathbf{e}}_\theta$ components of this flux and performing the appropriate average there results

$$\langle \hat{\mathbf{e}}_v \cdot \langle \hat{\Gamma}_{ql} \rangle_t \rangle_\phi = -\frac{q^2}{4m^2} \sum_{\text{rays}} \sum_n \mathbf{a}^* \cdot \left\{ \frac{\cos \theta}{\sin \theta} \hat{\mathbf{b}} J_n + \hat{\mathbf{e}}_- \, e^{-i\delta} J_{n+1} + \hat{\mathbf{e}}_+ \, e^{+i\delta} J_{n-1} \right\}^* \sin \theta$$

$$\times e^{-i(\chi - n\delta)} \int_{-\infty}^t d\tau \, e^{i(\chi' - n\delta')}$$

$$\qquad (3.1.94)$$

$$\times \mathbf{a}' \cdot \left\{ \hat{\mathbf{b}}' \frac{\cos \theta'}{\sin \theta'} J_n' \left[\sin \theta' \frac{\partial}{\partial v'} + \left(\frac{\cos \theta'}{v'} - \frac{1 - n\Omega'/\omega}{v' \cos \theta'} \right) \frac{\partial}{\partial \theta'} \right] \right.$$

$$+ [\hat{\mathbf{e}}'_+ \, e^{+i\delta'} J'_{n-1} + \hat{\mathbf{e}}'_- \, e^{-i\delta'} J'_{n+1}]$$

$$\left. \times \left[\sin \theta' \frac{\partial}{\partial v'} + \left(\frac{\cos \theta'}{v'} - \frac{k'_\parallel}{\omega} \right) \frac{\partial}{\partial \theta'} \right] \right\} \hat{f} + cc$$

and

$$\langle \hat{\mathbf{e}}_\theta \cdot \langle \hat{\Gamma}_{ql} \rangle_t \rangle_\phi = -\frac{q^2}{4m^2} \sum_{\text{rays}} \sum_n \mathbf{a}^* \cdot \left\{ \hat{\mathbf{b}} \frac{\cos \theta}{\sin \theta} J_n \left(\cos \theta - \frac{1 - n\Omega/\omega}{\cos \theta} \right) \right.$$

$$\left. + [\hat{\mathbf{e}}_- \, e^{-i\delta} J_{n+1} + \hat{\mathbf{e}}_+ \, e^{+i\delta} J_{n-1}] \left[\cos \theta - \frac{k_\parallel v}{\omega} \right] \right\}^*$$

$$\times e^{-i(\chi - n\delta)} \int_{-\infty}^t d\tau \, e^{i(\chi' - n\delta')} \qquad (3.1.95)$$

$$\times \mathbf{a}' \cdot \left\{ \hat{\mathbf{b}}' \frac{\cos \theta'}{\sin \theta'} J_n' \left[\sin \theta' \frac{\partial}{\partial v'} + \left(\frac{\cos \theta'}{v'} - \frac{1 - n\Omega'/\omega}{v' \cos \theta'} \right) \frac{\partial}{\partial \theta'} \right] \right.$$

$$+ [\hat{\mathbf{e}}'_+ \, e^{+i\delta'} J'_{n-1} + \hat{\mathbf{e}}'_- \, e^{-i\delta'} J'_{n+1}]$$

$$\left. \times \left[\sin \theta' \frac{\partial}{\partial v'} + \left(\frac{\cos \theta'}{v'} - \frac{k'_\parallel}{\omega} \right) \frac{\partial}{\partial \theta'} \right] \right\} \hat{f} + cc$$

Consider the trajectory integrals which appear in (3.1.94)–(3.1.95)

$$I = \int_{-\infty}^{t} d\tau \, \Pi(n, \tau) \, e^{i(\chi'(n,\tau) - n\delta')}. \tag{3.1.96}$$

For a given wave, the quantity δ' varies slowly on the magnetic field variation scale length. In a system periodic in s, following Bernstein and Baxter (1981), we resolve the integral (3.1.96) as a sum of integrals each constituting a single transit or bounce in the periodic field structure. Thus (3.1.96) can be written as the sum

$$I = \sum_{l=0}^{\infty} \int_{t-\tau_B}^{t} d\tau \, \Pi(n, \tau) \, e^{i(\chi'(n,\tau - l\tau_B) - n\delta')} \tag{3.1.97}$$

Since Π, δ are periodic (period τ_B), in view of (3.1.92) we can write

$$\chi(n, \tau - l\tau_B) = \chi(n, \tau) + l \int_{0}^{-\tau_B} d\tau \, v(n, \tau) = \chi(n, \tau) + l\chi_B(n). \tag{3.1.98}$$

Applying this result in (3.1.97), the trajectory integral is recast as the sum of integrals

$$I = \sum_{l=0}^{\infty} \int_{t-\tau_B}^{t} d\tau \, \Pi(n, \tau) \, e^{i(\chi'(n,\tau) - n\delta')} \, e^{il\chi_B}. \tag{3.1.99}$$

Upon summation, (3.1.99) becomes

$$I = \frac{1}{1 - e^{i\chi_B}} \int_{t-\tau_B}^{t} d\tau \, \Pi(n, \tau) \, e^{i(\chi'(n,\tau) - n\delta')} \tag{3.1.100}$$

Recognizing that the integrals appearing in (3.1.100) are simply bounce-averages and adopting the notation introduced in (3.1.52) the trajectory integral is recast as

$$I = \frac{\tau_B \langle\!\langle \Pi \, e^{i(\chi - n\delta)} \rangle\!\rangle}{1 - e^{i(\langle\!\langle v \rangle\!\rangle \tau_B)}} = \Upsilon \tau_B \langle\!\langle \Pi \, e^{i(\chi - n\delta)} \rangle\!\rangle. \tag{3.1.101}$$

The quantity referred to as Υ accounts for multiple bounce coherence effects. In actuality, processes such as collisional (extrinsic) or wave-particle interaction induced (intrinsic) gyro-phase diffusion will assure Υ remains close to unity. The quantity χ given by (3.1.92) is a representation of the expected advance of the interaction phase, v, along a given trajectory in a single bounce. This single bounce interaction phase is subject to stochastic variation due to collisions as well as due to the wave-particle interaction itself. Cohen *et al.* (1983) have shown the gyro-phase diffusion varies as t^3; this provides an effective means of washing-out multiple bounce resonances completely, and decorrelating single pass resonances to an extent roughly proportional to the cube of the number of gyro-periods separating them. In tokamaks the magnetic field connection length is quite long and in most cases of interest the plasma is sufficiently collisional to assure single bounce/transit effects are the only operant coherence effects of significance. These issues are considered in more detail in Appendix 3C.

Using (3.1.71), (3.1.94)–(3.1.95), and (3.1.100), a short calculation reveals the bounce-averaged quasilinear coefficients in a form compatible with (3.1.55): viz.

$$A_{0_{ql}} = D_{0_{ql}} = 0, \qquad (3.1.102)$$

$$B_{0_{ql}} = -\lambda \frac{q^2}{4m^2} \sum_{rays} \sum_n \left\{ \langle\!\langle \mathbf{a} \cdot \Theta_n v^2 \sin\theta\, e^{i\Psi} \rangle\!\rangle^* \Upsilon \tau_B \langle\!\langle \mathbf{a} \cdot \Theta_n \sin\theta\, e^{i\Psi} \rangle\!\rangle + cc. \right\}, \qquad (3.1.103)$$

$$C_{0_{ql}} = -\lambda \frac{q^2}{4m^2} \sum_{rays} \sum_n \left\{ \langle\!\langle \mathbf{a} \cdot \Theta_n v^2 \sin\theta\, e^{i\Psi} \rangle\!\rangle^* \Upsilon \tau_B \right.$$
$$\times \left\langle\!\!\left\langle \mathbf{a} \cdot \left[\Theta_n \left(\cos\theta - \frac{k_\parallel v}{\omega}\right) - \eta \hat{\mathbf{b}} \frac{\cos\theta}{\sin\theta} J_n \right] \right. \qquad (3.1.104)$$
$$\left. \times\, e^{i\Psi} \frac{1}{v}\frac{\partial\theta_0}{\partial\theta} \right\rangle\!\!\left\rangle + cc. \right\},$$

$$E_{0_{ql}} = -\lambda \frac{q^2}{4m^2} \sum_{rays} \sum_n \left\{ \left\langle\!\!\left\langle \frac{v\sin\theta}{\sqrt{\psi}} \frac{\partial\theta_0}{\partial\theta} \mathbf{a} \cdot \left[\Theta_n \left(\cos\theta - \frac{k_\parallel v}{\omega}\right) \right.\right.\right.$$
$$\left.\left.\left. - \eta \hat{\mathbf{b}} \frac{\cos\theta}{\sin\theta} J_n \right] e^{i\Psi} \right\rangle\!\!\right\rangle^* \Upsilon \tau_B \times \langle\!\langle \mathbf{a} \cdot \Theta_n \sin\theta\, e^{i\Psi} \rangle\!\rangle + cc. \right\}, \qquad (3.1.105)$$

$$F_{0_{ql}} = -\lambda \frac{q^2}{4m^2} \sum_{rays} \sum_n \left\{ \left\langle\!\!\left\langle \frac{v\sin\theta}{\sqrt{\psi}} \frac{\partial\theta_0}{\partial\theta} \mathbf{a} \cdot \left[\Theta_n \left(\cos\theta - \frac{k_\parallel v}{\omega}\right) \right.\right.\right.$$
$$\left.\left.\left. - \eta \hat{\mathbf{b}} \frac{\cos\theta}{\sin\theta} J_n \right] e^{i\Psi} \right\rangle\!\!\right\rangle^* \Upsilon \tau_B \times \left\langle\!\!\left\langle \mathbf{a} \cdot \left[\Theta_n \left(\cos\theta - \frac{k_\parallel v}{\omega}\right) \right.\right.\right.$$
$$\left.\left.\left. - \eta \hat{\mathbf{b}} \frac{\cos\theta}{\sin\theta} J_n \right] e^{i\Psi} \frac{1}{v}\frac{\partial\theta_0}{\partial\theta} \right\rangle\!\!\right\rangle + cc. \right\}. \qquad (3.1.106)$$

In the preceding, we have abbreviated $\Psi = \chi - n\delta$ and introduced η, Θ_n where

$$\eta = \frac{\omega - k_\parallel v \cos\theta - n\Omega}{\omega\cos\theta} = \frac{v}{\omega\cos\theta} \qquad (3.1.107)$$

vanishes at resonance, and

$$\Theta_n = \frac{\cos\theta}{\sin\theta} \hat{\mathbf{b}} J_n + \hat{\mathbf{e}}_- e^{-i\delta} J_{n+1} + \hat{\mathbf{e}}_+ e^{+i\delta} J_{n-1}. \qquad (3.1.108)$$

The bounce-averages in (3.1.103)–(3.1.106) may be evaluated by the method of stationary phase due to the rapid variation with τ of the exponent Ψ. The eikonal Ψ is expanded about the point where $d\Psi/dt$ vanishes (elsewhere the rapid oscillations provide canceling contributions to the integrand). By (3.1.92), this implies

$$\frac{d\Psi}{dt} = \frac{d}{dt}\int^t d\tau\, v(n, \tau) = v(n, t) = 0 \qquad (3.1.109)$$

the condition of wave-particle resonance.

In the small banana width approximation (vanishing $\mathbf{k}\cdot\mathbf{v}_{\text{drift}}$), η vanishes at the point of stationary phase, i.e., the wave-particle resonance. At points for which $dB/d\tau = 0$ coincidentally with resonances, not only is $v = 0$, but also $\dot{v} = 0$. The expansion of Ψ about these points must be extended to one higher order (including a term in $\ddot{v}$). For small $k_\parallel$, the same phenomena occur near orbit turning points. A detailed analysis of this procedure is presented in Appendix 3C. We can represent the phase integral in these cases mnemonically as

$$\int_0^{\tau_B} d\tau\, \Pi(n, \tau)\, e^{i\Psi} = \int_0^{\tau_B} d\tau\, \Pi \exp\left(-i\int_0^\tau d\tau'\,(\omega - k_\parallel v_\parallel - n\Omega)(\tau')\right)$$

$$= \int_0^{s_B} ds\, \Pi\tau_c\delta(s - s_{\text{res}}), \qquad (3.1.110)$$

where τ_c, which might be called an effective correlation time, is identified through evaluation of the trajectory integral; s_R is the position at which resonance occurs. Associated with the timelike quantity τ_c through the evaluation of Ψ at the point of stationary phase is a certain phase factor. It is the advance of this phase between subsequent resonances which determines the single bounce coherence of sequential wave-particle interactions. Orbit resonances which are too close to sense significant gyro-phase diffusion can interfere constructively or destructively through the action of the relative interaction phase separating them. Details of the calculation pertaining to this phenomenon are discussed in Appendix 3C.

In reality the wave-particle resonant interaction is not confined to a single point (s_R), but rather occurs over an interval corresponding roughly to the distance the particle (gyro-center) moves in a time of the order of $|\tau_c|$. We wish eventually to determine the spacial (poloidal) variation of RF power absorption so that the wave damping calculated through the quasilinear model can be used concurrently with a wave propagation calculation. To this end we introduce a resonance broadening model which, apart from a normalizing factor $|\tau_c|$, replaces the exponential in the integrand of (3.1.110) by a resonance weighting function $w(\tau)$ such that $\int d\tau\, w(\tau) = 1$, the weighting function is normalized; $\int d\tau\, \tau w(\tau) = \tau_R$, the weighting function is centered on the wave-particle resonance; and finally, $\int d\tau\, (\tau - \tau_R)^2 w(\tau) = |\tau_c|^2$, the width of the weighting function corresponds to the wave-particle correlation time. Clearly, the stationary phase method alone corresponds to a weighting function with point support, $w(\tau) = \delta(\tau - \tau_R)$.

This resonance broadening technique should be viewed more as a computational convenience than a theoretical generalization. We sacrifice no physical insight by neglecting the broadening in the description of the underlying structure of the theory. Thus, to simplify the discussion, let us restrict ourselves for the moment to the limiting case of point resonance. For finite $k_\parallel$, then, (3.1.110) is an adequate mnemonic representation everywhere except in the neighborhood of the tangent resonance points. At tangent resonance, while $v = 0$ and $\dot{v} = 0$, in general $v_\parallel \neq 0$, and there results from the trajectory integral a variant of the form (3.1.110) involving $d\Pi/ds$.

In the vicinity of tangent resonance a gyro-center trajectory passes through two resonances which are close to one another (in physical space). Since these resonances are close, there is a phase correlation between the two interactions which is, in general, not randomized. The role played by the Bessel functions in depicting wave-particle interaction interference (cf. (3.1.90), (3.1.91)) at the gyro-orbit level is, roughly speaking, played by the Airy function at the bounce-orbit level.

The method used to compute the quasilinear coefficients is outlined schematically in Figs. 3.1, 3.2, and 3.3, and detailed in Appendix 3C. Figure 3.1 shows the phaseflow corresponding to the set of velocity space meshpoints chosen to represent the distribution $\mathscr{F}$. The abscissa is the arclength along the magnetic field measured from the midplane normalized to the arclength from the midplane (minimum B) to the throat (maximum B). The ordinate is the cosine of the pitch-angle along a gyro-center trajectory. The orbits above the separatrix correspond to co-passing orbits, those below the separatrix correspond to counter-passing orbits, and the elliptical trajectories correspond to trapped orbits. Since kinetic energy is a constant of the motion, these orbits are independent of v_0. The vertical curves in Fig. 3.1 represent the relation $v = 0$ in this space for a set of values v_0. The solid curves are those for which $k_\parallel > 0$, and the dotted curves are those for which $k_\parallel < 0$. The intersections of the gyro-center trajectories and the curves $v = 0$ are the resonances corresponding to positions of stationary wave-particle interaction phase. For $k_\parallel = 0$, the vertical curves all collapse into a single vertical line passing through the common intersection of the curves. That intersection becomes the turning point or bounce-resonance point. For $k_\parallel \neq 0$ the loci of points in the phase space at which wave-particle resonance occurs coincidentally with $\dot{v} = 0$, tangent resonance, is shown as the remaining curve in Fig. 3.1. This curve, of course, also goes through the turning point resonance, since there is no parallel Doppler shift at that point.

Figure 3.2 shows the loci in the midplane velocity space of points corresponding to orbits all of which pass through resonance at the same arclength position (or equivalently, the same poloidal angle). Some velocity space points are covered by two resonance loci. These orbits pass through resonance twice, at two different arclength positions. Along the inner boundary of this region

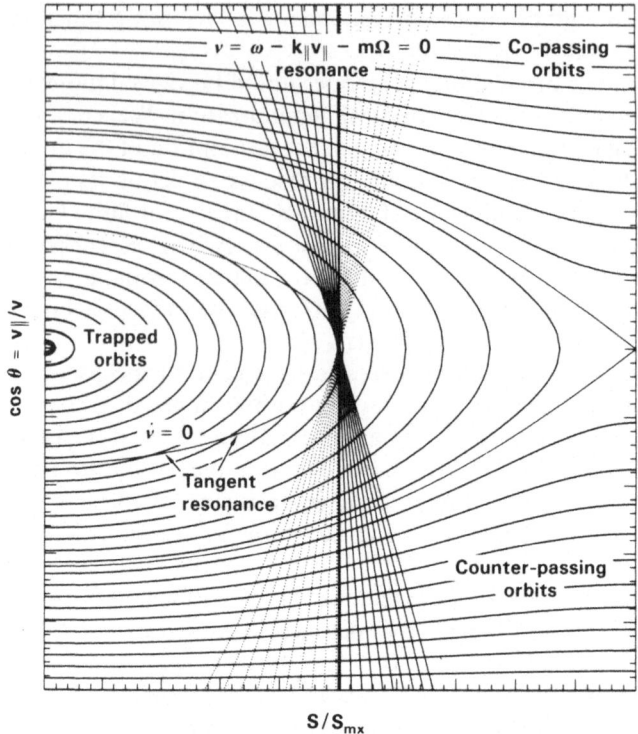

Figure 3.1. Phase flow corresponding to the set of velocity space meshpoints chosen to represent the distribution $\mathscr{F}$. The abscissa is the arclength along the magnetic field measured from the midplane normalized to the arclength from the midplane (minimum B) to the throat (maximum B). The ordinate is the cosine of the pitch-angle along a gyro-center trajectory. The orbits above the separatrix correspond to co-passing orbits, those below the separatrix correspond to counter-passing orbits, and the elliptical trajectories correspond to trapped orbits. Since kinetic energy is a constant of the motion (3.1.42), these orbits are independent of v_0. The vertical curves represent the relation $v = \omega - k_\parallel v_\parallel - m\Omega = 0$ in this space for a set of values v_0 on the chosen velocity mesh. The solid curves are those for which $k_\parallel > 0$ and the dotted curves are those for which $k_\parallel < 0$. The intersections of the gyro-center trajectories and the curves $v = 0$ are the resonances corresponding to positions of stationary wave-particle interaction phase. For $k_\parallel = 0$, the vertical curves all collapse into a single vertical line passing through the common intersection of the curves. That intersection corresponds to the turning point resonance. For $k_\parallel \neq 0$ the loci of points in the phase space at which wave–particle resonance occurs coincidentally with $\dot{v} = 0$, tangent resonance, is shown as the remaining curve. This curve, of course, also goes through the turning point resonance, since there is no parallel Doppler shift at that point.

the two resonance loci covering a given point are tangent to one another, and to a curve we will call the tangent resonance locus, shown more clearly in Fig. 3.3. As $k_\parallel \to 0$ the tangent resonance curves become straight lines. These radials coincide with the asymptotes of the family of quasilinear diffusion characteristics depicted in Fig. 3.4. Resonant diffusion for a definite wave-frequency is one dimensional. Phase space density diffuses along these curves under the action

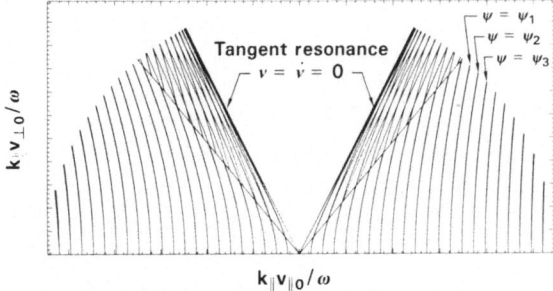

Figure 3.2. Resonance loci. Loci in the midplane velocity space of points corresponding to orbits all of which pass through resonance at the same arclength position (or equivalently, the same poloidal angle). Some velocity space points are covered by two resonance loci. These orbits pass through resonance twice, at two different arclength positions. Along the inner boundary of this region, the two resonance loci covering a given point are tangent to one another, and to a curve we will call the tangent resonance locus.

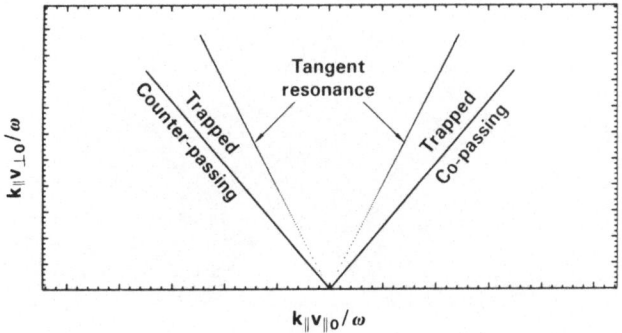

Figure 3.3. Tangent resonance loci. Loci in the midplane velocity space of points corresponding to orbits which pass through resonance ($v = \omega - k_\parallel v_\parallel - m\Omega = 0$) with $\dot{v} = 0$. As $k_\parallel \to 0$ the tangent resonance curves approach straight lines, corresponding to turning point resonance. These straight lines coincide with the quasilinear resonant diffusion asymptotes. Wave–particle interaction causes no pitch-angle diffusion along these lines.

of waves of a definite frequency, driven by density gradients locally tangent to the curves. This is in contrast to collisional diffusion which is two dimensional. The two-parameter family of collisional diffusion characteristics shown in Fig. 3.5 are just curves of constant energy and curves of constant pitch-angle. For small $k_\parallel$ the asymptotes in Fig. 3.4 correspond to orbits which bounce at resonance.

With the foregoing provisos, the quasilinear coefficients (3.1.103)–(3.1.106) can be cast in the mnemonic form

$$B_{0_{ql}} = \lambda S v^2 \sin^2 \theta \frac{\tau_c^2}{\tau_B}\bigg|_{res} , \qquad (3.1.111)$$

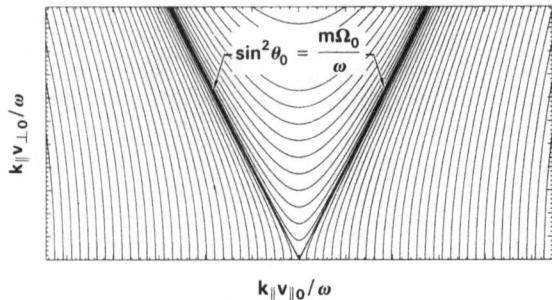

Figure 3.4. Diffusion characteristics of the quasilinear resonant diffusion operator. Resonant diffusion for a definite wave frequency is one dimensional. Phase space density diffuses along these curves under the action of waves of a definite frequency, driven by density gradients locally tangent to the curves. For small $k_\parallel$ the asymptotes correspond to orbits which bounce at resonance.

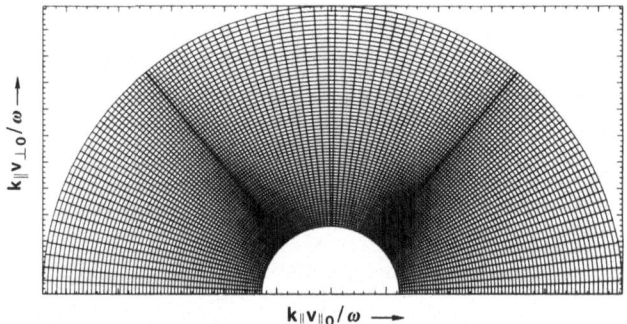

Figure 3.5. Collisional diffusion characteristics. This two-parameter family of diffusion characteristics coincides with the midplane velocity mesh. The curves are curves of constant energy (semicircles) and pitch-angle (radials). The nodes represent the meshpoints utilized in the calculations. Near the origin the mesh becomes too fine for photographic reproduction, hence the central white disc here and in some subsequent figures.

$$C_{0_{ql}} = \lambda S v \sin \theta \left[\cos \theta - \frac{k_\parallel v}{\omega} \right] \frac{\partial \theta_0}{\partial \theta} \frac{\tau_c^2}{\tau_B} \Bigg|_{res}, \qquad (3.1.112)$$

$$E_{0_{ql}} = \lambda S \frac{v}{\sqrt{\psi}} \sin^2 \theta \left[\cos \theta - \frac{k_\parallel v}{\omega} \right] \frac{\partial \theta_0}{\partial \theta} \frac{\tau_c^2}{\tau_B} \Bigg|_{res}, \qquad (3.1.113)$$

$$F_{0_{ql}} = \lambda S \frac{\sin \theta}{\sqrt{\psi}} \left[\cos \theta - \frac{k_\parallel v}{\omega} \right]^2 \left[\frac{\partial \theta_0}{\partial \theta} \right]^2 \frac{\tau_c^2}{\tau_B} \Bigg|_{res}, \qquad (3.1.114)$$

where λ was defined in (3.1.55), the quantity S is given by

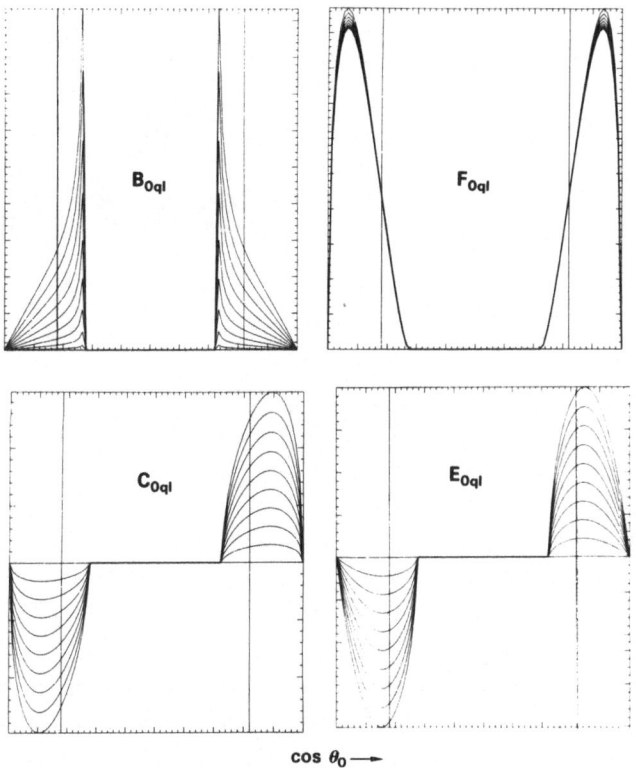

Figure 3.6. The quasilinear coefficients of (3.1.111)–(3.1.114) as functions of $\cos \theta_0$ for various values of v_0. The prominence in the energy diffusion coefficients occurs in the vicinity of tangent resonance. This particular case represents single frequency fundamental minority species excitation with rather small $k_\parallel$. The verticle lines coincide with the trapped/passing boundary. The energies represented are too small to see significant finite gyro-phase interfence effects, and the mesh spacing is too coarse to depict bounce-phase interference.

$$S = -\Upsilon \frac{q^2}{4m^2} \sum_{\text{rays}} \sum_n (\mathbf{a} \cdot \Theta_n)^* (\mathbf{a} \cdot \Theta_n), \qquad (3.1.115)$$

and all quantities are to be evaluated at the appropriate resonance(s) as determined through (3.1.110).

The quasilinear coefficients of (3.1.111)–(3.1.114) are shown in Fig. 3.6, where they are displayed as functions of $\cos \theta_0$ for various values of v_0. The prominence in the energy diffusion coefficients occurs in the vicinity of tangent resonance. This particular case represents single frequency fundamental minority species excitation with rather small $k_\parallel$. The vertical lines coincide with the trapped/passing boundary. The energies represented are too small to

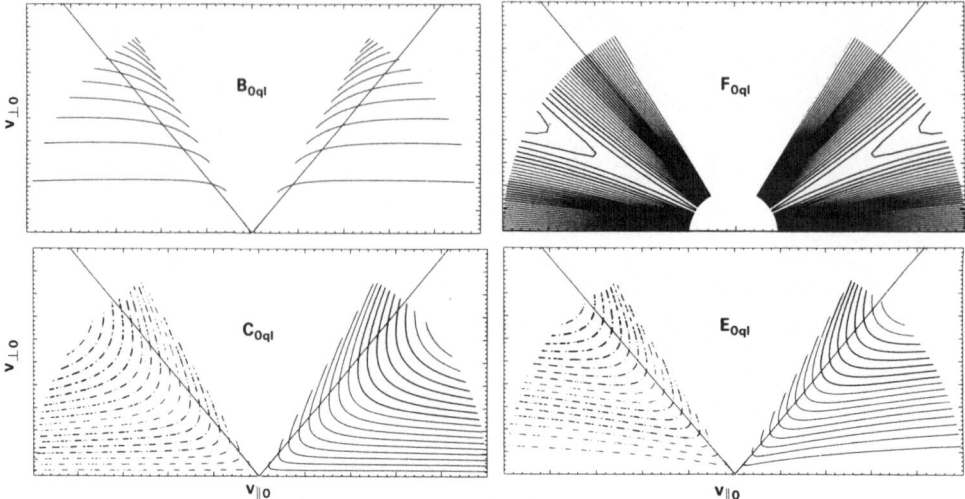

Figure 3.7. The same coefficients as Fig. 3.6 as a contour map in the midplane velocity space. There is no resonant diffusion in a wedge about $\pi/2$ since these trapped orbits do not transit through resonance (cf. Fig. 3.1).

see significant finite gyro-phase interference effects, and the mesh spacing is too coarse to depict any but the first bounce-phase interference fringe. Figure 3.7 shows the same coefficients as a contour map in the midplane velocity space.

Figure 3.8 shows the ion distribution function of a minority species excited at its fundamental cyclotron harmonic. Figure 3.9 shows the velocity space fluxes driven by: (a) the resonant diffusion; (b) the collisional diffusion; and (c) a combination of these. Notice the energy diffusion is maximal in the neighborhood of the tangent resonances, as described previously.

3.1.2.1. RF power absorption

The power absorbed on a flux surface per unit toroidal extent by resonant particles can be represented in terms of the quasilinear component of the

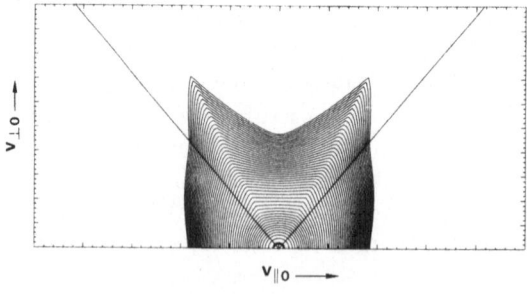

Figure 3.8. The ion distribution function of a minority species excited at its fundamental cyclotron harmonic at $t = 2$ ms.

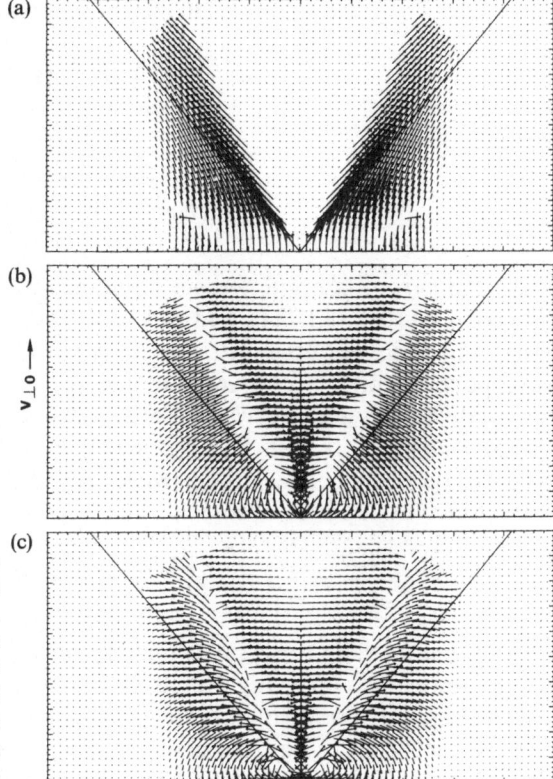

Figure 3.9. Velocity space fluxes for a minority species excited at its fundamental cyclotron harmonic. The three contour plots show the fluxes driven by: (a) resonant diffusion; (b) collisional diffusion; and (c) the combination of these. The energy diffusion is maximal in the neighborhood of the tangent resonances (cf. Appendix 3C).

Fokker–Planck operator

$$P = \int d^3 v_0 \tfrac{1}{2} m v_0^2 \frac{\partial \lambda \mathscr{F}_0}{\partial t} = - \int d^3 v_0 \tfrac{1}{2} m v_0^2 \nabla_{v_0} \cdot \hat{\Gamma}_{0_{ql}}.$$

Using the expression for the divergence of the flux in the form (3.1.55), performing an integration by parts, and recognizing that

$$C_{0_{ql}} = \frac{m\Omega_0/\omega - \sin^2 \theta_0}{v_0 \sin \theta_0 \cos \theta_0} B_{0_{ql}},$$

the RF power absorbed can be cast in the form

$$P_{\mathrm{RF}} = -2\pi \int dv_0 \int d\theta_0 \, m v_0 \sin \theta_0 B_{0_{ql}} \hat{\Delta}_0 \mathscr{F}_0, \qquad (3.1.116)$$

where the differential operator $\hat{\Delta}_0$ is given by

$$\hat{\Delta}_0 = \frac{\partial}{\partial v_0} + \frac{m\Omega_0/\omega - \sin^2 \theta_0}{v_0 \sin \theta_0 \cos \theta_0} \frac{\partial}{\partial \theta_0}.$$

Following the discussion of resonance broadening we set

$$B_{0_{ql}} = \bar{b}_0 \int d\tau \, w(\tau) \mathbf{E}^2(\tau),$$

or, equivalently,

$$B_{0_{ql}} = \bar{b}_0 \int d\theta_p \, W(\theta_p) \mathbf{E}^2(\theta_p),$$

where the gyro-phase interference contained in $\bar{b}_0$ can be identified through (3.1.103), (3.1.108), (3.1.111), and (3.1.115). The quantities $\bar{b}_0$ and W are functions of orbit invariants v_0, θ_0 as well as wave field parameters ω, $k_{\parallel}$ $k_{\perp}$ and wave polarization at resonance. In the present context, however, the field amplitude $\mathbf{E}^2$ for a given wave mode, is to be viewed as depending on the spacial variable θ_p as well. This is done for purposes of distributing the power exchange spacially between the particles and the wave field.

The differential power absorption can now be represented as the integral

$$\frac{dP_{\mathrm{RF}}}{d\theta_p} = -2\pi \int dv_0 \int d\theta_0 \, m v_0 \sin \theta_0 \bar{b}_0 W \mathbf{E}^2 \hat{\Delta}_0 \mathscr{F}_0. \qquad (3.1.117)$$

For finite $k_{\parallel}$ there arise situations in which ions can transit through two distinct resonances on a single (quarter) bounce or transit in the magnetic well. For well-separated resonances far from tangent resonance, the power exchanged between the particles and the field must be divided between the resonances. The absolute interaction phase (the gyro-phase at θ_p, say) is randomly distributed over the particle (gyro-) ensemble; the appropriate average must be performed to eliminate any such dependence. The apportionment can depend only on the uncorrelated local interaction strengths: whereas a single particle on any given bounce may absorb wave energy at one resonance and emit wave energy at another, the ensemble of particles on the same orbit will either absorb or emit as a whole, *but not both*. The net resulting power exchange will be distributed in θ_p according to the interaction strength associated with the local wave-particle resonance, i.e., $\bar{b}_0 W \mathbf{E}^2$.

A word about applications: The model as set out in the foregoing is presently incorporated in the CQL family of codes. The philosophy with which the present version was developed centers on the mechanism of magnetic decorrelation. Though a charged particle must be in resonance to interact strongly with the exciting RF fields, it is not the case that the magnetic correlation time is the most rational measure of the strength of the interaction in all cases. The magnetic correlation time does, however, represent a persistent uniform underlying structure. In some cases of interest, finite bandwidth wave spectra give rise to wave-particle correlation times shorter than τ_c. Models which combine these effects in a rational manner are the subject of current study.

For ICRH studies, we adopt a simplified approach in specifying the field amplitudes **a**. For each *wave* the parallel index of refraction *at resonance* is specified. The cold plasma dispersion relation can be used to compute the perpendicular index of refraction and the relative amplitudes of the right- and left-circularly polarized components. Alternatively, rays can be traced to their source where their amplitude is specified. For details of the various wave models incorporated in the source the reader is referred to Appendix 3D.

3.1.3. Velocity space boundary conditions

The peripheral boundary conditions enforced in the calculation are essentially identical to those in the two-dimensional uniform field calculation of Chapter 2 given in Section 2.1.3.2, i.e.,

$$\mathscr{F}_0(v_0 = 0, \theta_0) \text{ is independent of } \theta,$$

$$\frac{\partial \mathscr{F}_0}{\partial v_0}\left(v_0 = 0, \theta_0 = \frac{\pi}{2}\right) = 0,$$

$$\mathscr{F}_0(v_0 = \infty, \theta_0) = 0. \tag{3.1.118}$$

$$\frac{\partial \mathscr{F}_0}{\partial \theta_0}(v_0, \theta_0 = 0) = \frac{\partial \mathscr{F}_0}{\partial \theta_0}(v_0, \theta_0 = \pi) = 0, \quad \text{and}$$

$$\frac{\partial \mathscr{F}_0}{\partial \theta_0}\left(v_0, \theta_0 = \frac{\pi}{2}\right) = 0.$$

This last equation reflects the requirement that the bounce-averaged distribution be symmetric about $\pi/2$ in the trapped orbit region. This simply reiterates the fact that $\mathscr{F}$ is not an explicit function of s, and the two points in question signify the same orbit at ϕ_B and $\phi_B + \pi$.

There arises a complication at the trapped/passing boundary which does not appear in the uniform field theory: the bounce time diverges logarithmically there and the formal ordering (3.1.8) fails. Thus a valid set of jump conditions is required in the vicinity of the boundary layer separating trapped- and circulating-particle orbits in order to properly couple the solutions in the various neighboring regions. From (3.1.55) it is apparent that to satisfy conservation of particles (3.1.54), in addition to continuity of $\mathscr{F}_0$ at the trapped/ passing boundary, we must also require the normal flux of $\lambda\mathscr{F}_0$ to satisfy

$$0 = [[\hat{\Gamma}_0 \cdot \hat{\mathbf{e}}_{\theta_0}]] = \left[\left[\left(D_0 + E_0\frac{\partial}{\partial v_0} + F_0\frac{\partial}{\partial \theta_0}\right)\mathscr{F}_0\right]\right] = [[\mathscr{H}_0]]. \tag{3.1.119}$$

Here the [[ ]] means the jump (in the flux) between leaving the trapped region and entering the passing region. We note in passing that the orbit topology determines the form of the jump condition in its application (see Section 3.2.3). Here three distinct regions of velocity space are in contact; co-passing, counter-passing, and trapped orbits. The jump condition is merely a statement of the fact that all trapped particles which become passing particles, by collisions for example, must cross over the trapped/passing boundary and appear on the other side.

While the flux of $\lambda \mathscr{F}_0$ is not defined on the boundary, it is well defined to each side of the boundary. We must require that $\mathscr{F}_0$ be continuous but there may be a discontinuity in its normal derivative. Furthermore, discontinuity in the coefficients may arise as a consequence of the RF excitation.

The procedure by which (3.1.119) is implemented is described in detail in Appendix 3B and Section 3.2.4.

3.1.4. Particle and energy source and loss terms

In addition to the RF heating term described separately in Section 3.1.2, there are three other source and loss mechanisms considered in this model, the DC electric field, fusion reactions, and neutral beam sources.

The local gyro-phase averaged kinetic equation described by (3.1.33) contains a term $q/m(\mathbf{E} \cdot \hat{\mathbf{b}}) \, \partial f/\partial v_{\parallel}$. Expanding f as in (3.1.45) and rewriting in (v_0, θ_0) coordinates we obtain

$$\frac{q}{m} \mathbf{E} \cdot \hat{\mathbf{b}} \frac{\partial \mathscr{F}}{\partial v_{\parallel}} = \frac{q}{m} E_{\parallel} \cos \theta \left(\frac{\partial}{\partial v_0} - \frac{\sin \theta_0}{v_0 \cos \theta_0} \frac{\partial}{\partial \theta_0} \right) \mathscr{F}, \qquad (3.1.120)$$

where terms with vanishing bounce-average have been suppressed. Treating the parallel electric field as perturbative and bounce-averaging this term yields

$$\left\langle\!\!\left\langle \frac{q}{m} \mathbf{E} \cdot \hat{\mathbf{b}} \frac{\partial f}{\partial v_{\parallel}} \right\rangle\!\!\right\rangle = -s^* E_{\parallel 0} \frac{q}{m} \frac{\sigma(\theta_0)}{v_0 \tau_B(\theta_0)} \left(\frac{\partial}{\partial v_0} - \frac{\sin \theta_0}{v_0 \cos \theta_0} \frac{\partial}{\partial \theta_0} \right) \mathscr{F}_0(v_0, \theta_0)$$
$$(3.1.121)$$

in the passing region and zero in the trapped region. The factor $\sigma(\theta_0)$ is a multiplier equal to 1 if $\theta_0 < \pi/2$ and equal to -1 if $\theta_0 > \pi/2$. The symbol s^* denotes the integral $\int ds \, \psi$. The additions to the coefficients A_0 (3.1.71) and D_0 (3.1.71) necessary to represent the effects of a DC ohmic electric field are

$$A_{E0} = -s^* E_{\parallel 0} \frac{q}{m} v_0^2 \cos \theta_0 \qquad (3.1.122)$$

and

$$D_{E0} = s^* E_{\parallel 0} \frac{q}{m} v_0 \sin^2 \theta_0. \qquad (3.1.123)$$

The fusion reaction loss term and the neutral beam source term are mathematically similar and are expressed in terms of their cross section profiles. We add a term $\lambda \langle\!\langle S_a(v, \theta, s) \rangle\!\rangle$ to the right-hand side of (3.1.55) obtaining

$$\frac{\partial \lambda \mathscr{F}_0}{\partial t} = \gamma \left(\frac{1}{v_0^2} \frac{\partial \mathscr{G}_0}{\partial v_0} + \frac{1}{v_0^2 \sin \theta_0} \frac{\partial \mathscr{H}_0}{\partial \theta_0} \right) + \lambda \langle\!\langle S_a(v, \theta, s) \rangle\!\rangle, \quad (3.1.124)$$

where

$$S_a(v, \theta, s) = S_a^n(v, \theta, s) \left\{ \sum_b \int dv' \, \sigma_i^{ab}(|\mathbf{v} - \mathbf{v}'|)|\mathbf{v} - \mathbf{v}'|\mathscr{F}_b(v', \theta', s) \right.$$

$$\left. + \sum_b \int dv' \, \sigma_{cx}^{ab}(|\mathbf{v} - \mathbf{v}'|)|\mathbf{v} - \mathbf{v}'|\mathscr{F}_b(v', \theta', s) \right\}$$

$$- \mathscr{F}_a(v, \theta, s) \sum_b \int dv' \, \sigma_{cx}^{ab}(|\mathbf{v} - \mathbf{v}'|)|\mathbf{v} - \mathbf{v}'|S_b^n(v', \theta', s)$$

$$- \mathscr{F}_a(v, \theta, s) \int dv' \, \sigma_{fus}^{ab}(|\mathbf{v} - \mathbf{v}'|)|\mathbf{v} - \mathbf{v}'|\mathscr{F}_b(v', \theta', s). \quad (3.1.125)$$

Here σ_i^{ab} is the relevant ionization cross section of species "a" by species "b", σ_{cx}^{ab} is the relevant charge-exchange cross section and σ_{fus}^{ab} is the fusion cross section. The quantity $S_a^n(v, \theta, s)$ is the neutral beam profile of species "a" at arclength s from the midplane, an input quantity, and $\mathscr{F}_b(v, \theta, s)$ is the bounce-averaged distribution function at arclength s from the midplane.

Since these three-dimensional integrals must be evaluated frequently (at every timestep, at every velocity meshpoint, and at each point along the orbit) the calculation must proceed expeditiously. The procedure involves a transformation of coordinates similar to that employed by Cordey et al. (1978). Consider a typical integral

$$R(v, \theta, s) = \int dv' \, \sigma(|\mathbf{v} - \mathbf{v}'|)|\mathbf{v} - \mathbf{v}'|\tau(v', \theta', s). \quad (3.1.126)$$

This may be rewritten as

$$R(v, \theta, s) = \int_0^\infty dv' \, v'^2 \int_0^\pi d\theta_{12} \sin \theta_{12} \int_0^{2\pi} d\phi_{12} \, \tau(v', \theta', s)\sigma(u)u, \quad (3.1.127)$$

where $u = |\mathbf{v} - \mathbf{v}'|$. The variables v', θ_{12}, and ϕ_{12} are coordinates in a spherical coordinate system in which the polar axis is along $\mathbf{v}$. ϕ_{12} is measured from the plane which includes $\mathbf{v}$ and the polar axis in the original (v, θ, ϕ) coordinate system. θ_{12} is the angle between $\mathbf{v}$ and $\mathbf{v}'$. Further, it can be shown that

$$\cos \theta' = \cos \theta \cos \theta_{12} + \sin \theta \sin \theta_{12} \cos \phi_{12}, \quad (3.1.128)$$

and

$$u = |\mathbf{v} - \mathbf{v}'| = \sqrt{v^2 + v'^2 - 2vv' \cos \theta_{12}}). \quad (3.1.129)$$

Expanding τ in a Legendre series, one obtains

$$\tau(v', \theta', s) = \sum_{l=0}^\infty \hat{t}_l(v', s)P_l(\cos \theta'), \quad (3.1.130)$$

with

$$\hat{t}_l(v', s) = \frac{2l + 1}{2} \int_0^\pi d\theta' \sin \theta' P_l(\cos \theta') \tau(v', \theta', s). \tag{3.1.131}$$

This may be inserted into (3.1.127) to obtain

$$R(v, \theta, s) = \int_0^\infty dv' \, v'^2 \int_0^\pi d\theta_{12} \sin \theta_{12} \int_0^{2\pi} d\phi_{12} \sum_{l=0}^\infty \hat{t}_l(v', s) P_l(\cos \theta') \sigma(u) u. \tag{3.1.132}$$

Using the relationship

$$P_l(\cos \theta') = \sum_{m=-l}^{m=+l} \frac{(l - m)!}{(l + m)!} P_l^m(\cos \theta) P_l^m(\cos \theta_{12}) \cos(m\phi_{12}), \tag{3.1.133}$$

substituting back into (3.1.132), and noting that the integral over ϕ_{12} is zero except for the case $m = 0$ ($P_l^0 = P_l$), there results

$$R(v, \theta, s) = 2\pi \sum_{l=0}^\infty P_l(\cos \theta) \int_0^\infty dv' \, v'^2 \hat{t}_l(v', s) \int_0^\pi d\theta_{12} \sin \theta_{12} P_l(\cos \theta_{12}) \sigma(u) u. \tag{3.1.134}$$

The utility of this representation is seen by noting: (a) the integral over θ_{12} is time-independent, and need be evaluated only once; (b) the sum over the Fourier–Legendre coefficients is independent of v whereas the integral over v' is independent of θ.

With the representation (3.1.134) it becomes possible to evaluate the local source (3.1.125). With the standard bounce-averaging procedure applied to the local source there obtains $\lambda \langle\!\langle S(v, \theta, s) \rangle\!\rangle$.

Furthermore (defining $\dot{n}(s) = \int d^3v \, S(v, \theta, s)$, the local particle source) the total time rate of change of the number of particles in the ensemble due to sources is given by

$$\dot{N} = \int_0^{s_{mx}} ds \frac{\dot{n}(s)}{\psi(s)} = \int d^3v_0 \, \lambda \langle\!\langle S(v, \theta, s) \rangle\!\rangle (v_0, \theta_0). \tag{3.1.135}$$

Each of these integrals is evaluated for the sake of comparison, as a diagnositic measure. Generally agreement is within 2%, except in cases in which highly anisotropic (in θ) sources are used, indicating additional resolution in the θ_0 mesh is necessary.

3.1.5. Velocity space loss region models

It is often the case in magnetic fusion devices that various classes of charged-particle orbits are not confined by the magnetic field structure; these orbits are called loss orbits. In the case of magnetic mirror traps, loss orbits are typically associated with small pitch-angle and gyro-centers can freely stream out of the trap along field lines in a time of the order of a transit time. In toroidal devices field lines are generally confined to a limited volume and gyro-centers cannot leave the confinement region by following field lines for a single transit time. However, since the field lines are necessarily curved, there arise cross-field drifts which carry gyro-centers across flux surfaces and, in some cases, out of the region of useful confinement into a wall, limiter, gas layer, or divertor. Two loss orbit models are implemented in the CQL code family to suggest the physical behavior of plasmas confined in devices in which there exist such loss regions.

3.1.5.1. *Mirror orbit loss model*

A time-invariant square-well electrostatic potential model descriptive of tandem mirror central cells is assumed; ion orbits passing through the potential jump at the mirror throat are lost from the system. A class of magnetically untrapped orbits are reflected by the potential jump and are thus electrostatically trapped.

A boundary layer arises in the vicinity of the loss boundary due to the various diffusive processes driving particles onto loss orbits in competition with the finite loss time. The relative rates of these competing processes determines the density of the layer.

To find a representation for the loss boundary in the midplane velocity coordinate system we set the parallel kinetic energy at the throat equal to the potential jump. All those particles whose parallel energy at the throat (point of maximum $|B|$) exceeds the potential jump are lost

$$\frac{mv_\parallel^2}{2}(s_{mx}) = e\Phi. \tag{3.1.136}$$

By energy conservation

$$v^2 = v_\parallel^2 + v_\perp^2 = v_\parallel^2 + \frac{2\mu B_0 \psi}{m} = v_{\parallel 0}^2 + v_{\perp 0}^2. \tag{3.1.137}$$

Solving for $v_\parallel$ and substituting in (3.1.136) yields

$$\frac{mv_\parallel^2}{2}(s_{mx}) = e\Phi = \frac{m}{2}\left(v^2 - \frac{2\mu B_0 \psi_{mx}}{m}\right). \tag{3.1.138}$$

This expression can be recast in the form

$$v_{\parallel 0}^2 - v_{\perp 0}^2(\psi_{mx} - 1) = \frac{2e\Phi}{m}, \tag{3.1.139}$$

hyperbolas which represent the midplane loss boundary for throat potential jump of magnitude Φ.

3.1.5.2. *Toroidal orbit loss model*

The toroidal loss model is used for tokamak calculations. Toroidal axisymmetry induces the conservation of toroidal canonical angular momentum. This fact is exploited to calculate the radial excursion from the bounce-averaged (or zero banana width) flux surface to first order in the parameter ρ_p/a, assumed small throughout. Here ρ_p is the gyro-radius in the poloidal magnetic field and a is the minor radius of the plasma.

The interaction of the particle along its perturbed orbit with neutral gas near the wall or limiter structures is computed and used to establish the probability that the particle is lost from its (unperturbed) orbit on any given bounce. A particle whose excursion brings it into contact with the wall is lost in a bounce time with probability unity.

The toroidal canonical angular momentum is given to requisite order by

$$P_\zeta = mv_\parallel R\hat{\mathbf{b}}\cdot\hat{\mathbf{e}}_\zeta + \frac{q}{c}\Psi, \tag{3.1.140}$$

where

$$\Psi = \Psi_0 - R_0\int_0^r \frac{dr'}{r'}\int_0^{r'} dr''\, r'' j_\zeta(r'') = \int_r^a dr\, RB_\theta \tag{3.1.141}$$

and j_ζ is the toroidal electrical current density.

Since P_ζ is a constant of the motion, upon bounce-averaging (3.1.140) we have to first order in ρ_p/a.

$$\langle\!\langle P_\zeta\rangle\!\rangle = P_\zeta = mv_\parallel R\hat{\mathbf{b}}\cdot\hat{\mathbf{e}}_\zeta + \frac{q}{c}\Psi = m\langle\!\langle v_\parallel R\rangle\!\rangle\hat{\mathbf{b}}\cdot\hat{\mathbf{e}}_\zeta + \frac{q}{c}\langle\!\langle\Psi\rangle\!\rangle, \tag{3.1.142}$$

which defines the bounce-averaged flux as

$$\langle\!\langle\Psi\rangle\!\rangle = \Psi + \frac{mc}{q}\hat{\mathbf{b}}\cdot\hat{\mathbf{e}}_\zeta(v_\parallel R - \langle\!\langle v_\parallel R\rangle\!\rangle). \tag{3.1.143}$$

We wish to identify $\langle\!\langle\Psi\rangle\!\rangle$ with the flux surface on which the zero banana width calculation takes place (zeroth order in ρ_p/a). This then implies that

$$\Psi = \langle\!\langle\Psi\rangle\!\rangle \tag{3.1.144}$$

and (as a direct result, regarding (3.1.142)) that

$$\langle\!\langle v_\parallel R\rangle\!\rangle = v_\parallel R. \tag{3.1.145}$$

Now for a trapped particle, to first order

$$\langle\!\langle v_\parallel R\rangle\!\rangle = 0 \tag{3.1.146}$$

so that the bounce-average of Ψ for a trapped particle is just the value of Ψ at the banana tips or turning points.

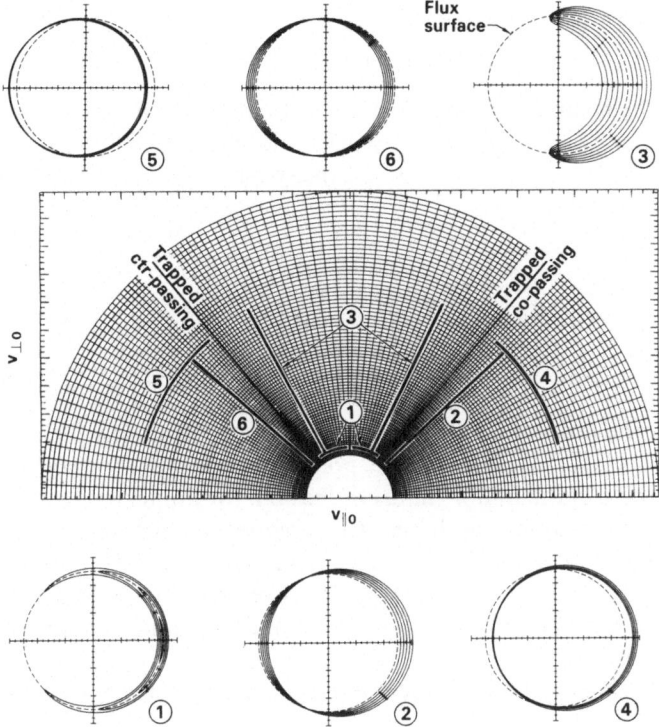

Figure 3.10. Gyro-center orbits projected on a constant toroidal angle plane for various energies and various pitch-angles as indicated.

We proceed to use (3.1.142) in conjunction with (3.1.141) to calculate the radial excursion $\Delta(r/a)$ as a function of the bounce-phase for a (trapped) orbit characterized by the initial conditions $\Psi(r/a) = \langle\!\langle\Psi\rangle\!\rangle$ at $v_{\parallel} = 0$. Figure 3.10 shows the result of this calculation for a set of orbits of increasing energy. The banana tips are all located on the same flux surface, $\langle\!\langle\Psi\rangle\!\rangle$. Having thus calculated the perturbed orbit (to lowest significant order) we can compute the probability that the particle survives a bounce by performing the attenuation integral

$$P = \exp\left(-\int dt/\tau(t)\right), \tag{3.1.147}$$

where $\tau = (n\langle\sigma v\rangle)^{-1}(t)$ is the mean interaction time along the orbit parametrized by t. For example, one might model the interaction with a neutral gas layer of specified properties and consider charge exchange as the principal extinction process. The result will be a bounce-averaged Krook operator. It is a short exercise to recast (3.1.147) as

$$P = \exp(-v_0 \tau_B \langle\langle 1/\lambda \rangle\rangle), \tag{3.1.148}$$

where λ is the mean free path associated with interaction time τ. For simplicity, let us represent λ as a constant $\lambda = 1/n\sigma$ within the gas layer, zero within the wall, and infinite where the neutral density vanishes. Equation (3.1.148) may then be written

$$P = \begin{cases} 1, & r(s_B) < r_G, \\ \exp\left(-\dfrac{n\sigma}{2}\displaystyle\int_{s_G}^{s_B} ds/\cos\theta\right), & r_G < r(s_B) < a, \\ 0, & r(s_B) > a, \end{cases} \tag{3.1.149}$$

where s_G is the arclength along the guiding center orbit at which the paraticle enters the gas layer, s_B is the bounce point, r_G is the inner (minor) radius of the gas layer, and a is the minor radius of the wall (or limiter).

A similar procedure may be used to account for circulating particles with the exception that (3.1.146) no longer holds and (3.1.145) must be solved to evaluate $\Psi(\phi_B)$, or alternatively $r(\phi_B)$ by inverting $\Psi(r) = \Psi(\phi_B)$. Figure 3.10 also shows a series of (projected) co-passing and counter-passing orbits for a set of energies.

The probability (3.1.149) is then used (implicitly) to form a Krook operator of the form

$$S = -\frac{1 - P}{\tau_B}, \tag{3.1.150}$$

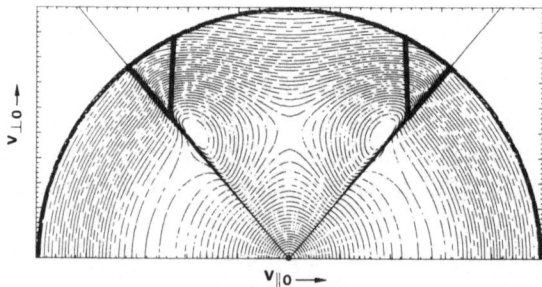

Figure 3.11. Contour map of the bounce-averaged charge–exchange Krook term on midplane velocity space is discussed in Sections 3.1.5 and 3.3.2. Contours are level curves of ion loss by charge exchange with the ambient neutral background resulting from a neutral transport calculation. The solid bold contours show where the edge of the gas layer and the wall would be, in the sense that orbits on these contours have finite banana excursions which bring them into contact with those respective (minor) radii ($r/a = 0.9$, 1.0). The position of the flux surface with which the (small banana width) ensemble is associated is $r/a = 0.85$.

which acts (on $\mathscr{F}_0$) as a sink of particles in the appropriate regions of velocity space; i.e., where particle orbits transit (radially) through the interaction region or hit the wall.

A more elaborate Krook term involving the results of the neutral transport calculation is discussed further in Section 3.3.2. Figure 3.11 represents a contour map of that calculation. The solid bold contours show where the edge of the gas layer and the wall would be, in the sense that orbits on these contours have finite banana excursions which bring them into contact with the respective (minor) radii.

3.2. Numerical Solution of Bounce-Averaged Fokker–Planck Equations

The procedures, techniques, and arrays employed in the bounce-averaged code family CQL are generally similar to those used in their precursor family FPPAC designed to solve Fokker–Planck equations in a uniform magnetic field, and described in Chapter 2. Cutler *et al.* (1977), and Matsuda and Stewart (1984) have also developed bounce-averaged Fokker–Planck codes, though each with an approach different from that discussed here. The Cutler code is a finite difference code primarily designed to study mirror confined plasma. The Matsuda/Stewart code is a finite element code designed for a more general multiregion problem and equipped with an elliptic solver for evaluating the Rosenbluth potentials. The CQL codes are designed primarily for tokamaks in which there reside magnetically confined trapped and passing populations of particles. The principal differences between the two families of codes FPPAC and CQL are associated with the bounce-averaging procedure and the maintenance of certain symmetries required by the orbit topology. Naturally, the distribution functions must be symmetric in the trapped region of velocity space: there is no distinction to be drawn between the populations on the left-going and right-going legs of trapped orbit. The coefficients of the bounce-averaged Fokker–Planck operator must have the appropriate symmetries to maintain this property at every stage of the evolution of the bounce-averaged distributions. Goldston (1977) has developed a code designed to treat tokamak plasmas and that code has been modified to conservative form and generalized by Hammett *et al.* (1983).

3.2.1. Numerical methods

The first three parts of this section are associated with the calculation of the bounce-averaged coefficients; the fourth describes the methods employed to time advance the equation with particular attention to the peculiarities at the trapped/passing boundary.

In general, arrays and coefficients are defined in CQL to correspond to those in FPPAC given in Section 2.4.1. In particular, the θ_0 mesh is indexed by "i",

$$\{\theta_i, \quad i = 1, I\},$$

where $\theta_1 = 0$, $\theta_I = \pi$ and we have dropped the midplane designating subscript for notational convenience. Similarly, the velocity v_0 mesh is indexed by "j",

$$\{v_j, \quad j = 1, J\},$$

with $v_1 = 0$. In addition to this midplane velocity space mesh, evaluating orbit integrals necessitates the designation of an orbit mesh.

$$\{s_l, \quad l = 1, L\},$$

where s_l is the arclength along a particle orbit from the starting point in the midplane to the lth meshpoint, and $s_L = \pi q R_0 = s_{mx}$, defined in (3.1.53), is the arclength from the midplane to the field maximum, also referred to as the throat. We define an auxiliary array

$$\{l_{\max}(i), \quad i = 1, I\}$$

to mean the *largest* orbit mesh index "l" such that a bouncing particle with midplane coordinates (v_j, θ_i) attains and passes s_l. If a particle should bounce precisely at s_l, the choice $l_{\max}(i) = l - 1$ is made. For passing particles $l_{\max}(i) = L - 1$ since all such orbits arrive at and transit through the point $s = s_{mx}$.

The bounce-average of a function $g(v, \theta, s)$ as defined in (3.1.52) is then approximated by

$$
\begin{aligned}
\tau_B(v_j, \theta_i) \langle\!\langle g(v, \theta, s) \rangle\!\rangle \Big|_{v_j, \theta_j} &= \int_0^{s_B} \frac{ds}{v_j \cos \theta(\theta_i, s)} g(v_j, \theta(\theta_i, s), s) \\
&\cong \sum_{l=1}^{l_{\max}(i)+1} g_l \, d\tau_B \,(j, i, l) \\
&= g_1 \int_0^{\bar{s}_2} \frac{ds}{v_\parallel} \left[1 - \frac{\psi - \psi_1}{\psi_2 - \psi_1} \right] \\
&\quad + \sum_{l=2}^{l_{\max}(i)} g_l \left\{ \int_{s_{l-1}}^{s_l} \frac{ds}{v_\parallel} \left[\frac{\psi - \psi_{l-1}}{\psi_l - \psi_{l-1}} \right] \right. \\
&\quad \left. + \int_{s_l}^{\bar{s}_{l+1}} \frac{ds}{v_\parallel} \left[1 - \frac{\psi - \psi_l}{\psi_{l+1} - \psi_l} \right] \right\} \\
&\quad + g_{l_{\max}(i)+1} \int_{s_{l_{\max}(i)}}^{\bar{s}_{l_{\max}(i)+1}} \frac{ds}{v_\parallel} \left[\frac{\psi - \psi_{l_{\max}(i)}}{\psi_{l_{\max}(i)+1} - \psi_{l_{\max}(i)}} \right],
\end{aligned}
\tag{3.2.1}
$$

where

$$\bar{s}_l = \min\{s_l, s_B(\theta_i)\},$$

$$g_l = g(v_j, \theta(\theta_i, \bar{s}_l), \bar{s}_l),$$

$$\bar{\psi}_l = \psi(\bar{s}_l),$$

and s_B is defined through (3.1.53) and the bounce time on the left-hand side of (3.2.1) is given by the sum

$$\tau_B(v_j, \theta_i) = \sum_{l=1}^{l_{\max}(i)+1} d\tau_B(j, i, l).$$

This algorithm corresponds to an assumption that $g(v, \theta, s)$ is linear in ψ between meshpoints. In general, it is assumed that ψ is known as a function of s allowing for simple and highly accurate evaluation of $d\tau_B(j, i, l)$ in (3.2.1).

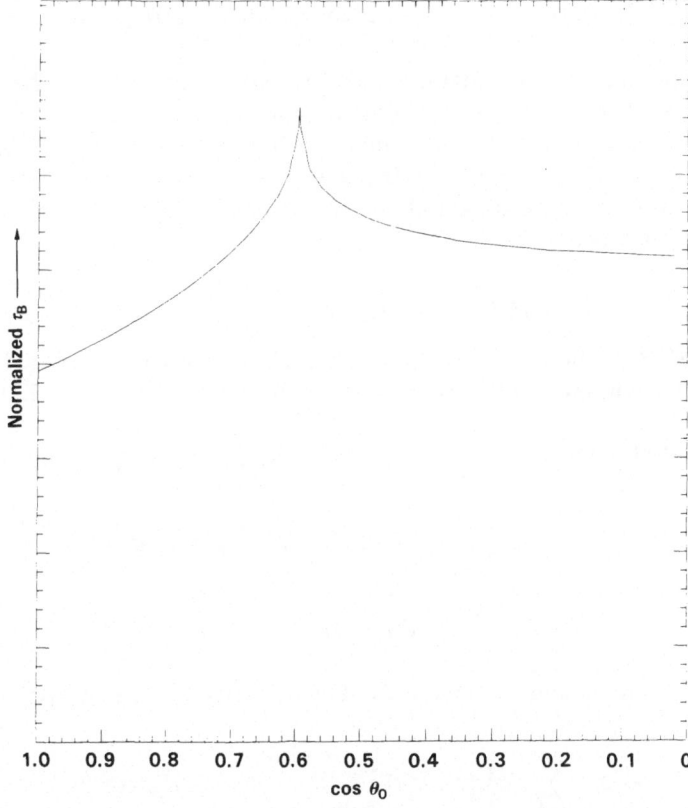

Figure 3.12. Bounce-time τ_B as a function of $\cos\theta_0$. The value of τ_B at the position of the logarithmic singularity is assigned by an averaging method detailed in Section 3.2.3 and Appendix 3B.

The result of the calculation is graphed in Fig. 3.12. Notice that the positon of the singularity in τ_B has been assigned a finite value. This value is determined by an averaging method described in Section 3.2.3 and Appendix 3B.

3.2.2. Bounce-averaging the Fokker–Planck coefficients

The procedure for evaluating the Fokker–Planck coefficients in an inhomogeneous magnetic field is similar to that used in the uniform field case locally in s. Once the local coefficients are determined, it is then necessary to combine the results with the proper weighting factors to generate the relevant bounce-averaged quantities. Since the Fokker–Planck coefficients must be evaluated not only in the midplane, but at each s_l meshpoint as well, an efficient numerical scheme must be devised to evaluate the Fourier–Legendre coefficients of the local distribution function (3.1.63),

$$V_m^b(v, s; t) = \frac{2m + 1}{2} \int_{-1}^{+1} d(\cos \theta) \mathscr{F}_b(v, \theta, s; t) P_m(\cos \theta). \qquad (3.2.2)$$

The evaluation of this integral is complicated by the fact that the distribution function $\mathscr{F}_b(v, \theta, s; t)$ is known only through $\mathscr{F}_{b0}(v_0, \theta_0; t)$ given by (3.1.47). While the orbit equations (3.1.41) and (3.1.42) may be employed to determine $\mathscr{F}_b$ at any point (v, θ). It is advisable from the point of view of computational efficiency and accuracy to avoid interpolation. To that end the following procedure is adopted.

Define

$$\cos z(i, l) = \cos \theta(\theta_i, s_l) \sqrt{1 - \psi_l \sin^2(\theta_i)}. \qquad (3.2.3)$$

Thus, $\cos z(i, l)$ is the cosine of the pitch-angle that a particle would have at orbit point s_l if it had a pitch-angle θ_i at the midplane. Splitting (3.2.2) as

$$V_m^b(v, s; t) = -\frac{2m + 1}{2} \left\{ \int_1^0 d(\cos \theta) \mathscr{F}_b(v, \theta, s; t) P_m(\cos \theta) \right.$$

$$\left. + \int_0^{-1} d(\cos \theta) \mathscr{F}_b(v, \theta, s; t) P_m(\cos \theta) \right\} \qquad (3.2.4)$$

$$= -\frac{2m + 1}{2}(V_L + V_U),$$

we can compute V_U and V_L separately. The quantity V_L may be approximated by

$$V_L \cong \sum_{i=1}^{i_{max}(l)-1} \mathscr{F}_0(v_j, \theta_i; t) \int_{\cos z(i, l)}^{\cos z(i+1, l)} d(\cos \theta) P_m(\cos \theta)$$

$$+ \mathscr{F}_0(v_j, \theta_{i_{max}(l)}; t) \int_{\cos z(i_{max}(l), l)}^0 d(\cos \theta) P_m(\cos \theta), \qquad (3.2.5)$$

where

$$\mathscr{F}_0(v_j, \theta_i; t) = \tfrac{1}{2}(\mathscr{F}_0(v_j, \theta_i; t) + \mathscr{F}_0(v_j, \theta_{i+1}; t))$$

and $i_{\max}(l)$ contains the maximum index "i" such that a particle with midplane coordinates $(v_0, \theta_0) = (v_j, \theta_i)$ will appear at orbit point s_l. Higher values of "i" correspond to particles that bounce before reaching s_l. There is a similar representation for V_U.

The procedure is streamlined by noting two points. First, since

$$\int_a^b d\mu \, P_m(\mu) = \frac{1}{2m + 1}(P_{m+1}|_a^b + P_{m-1}|_a^b) \tag{3.2.6}$$

the integrals in (3.2.5) can be computed analytically. Furthermore, since $\mathscr{F}_0$ is symmetric about $\tfrac{1}{2}\pi$ in the trapped particle orbit region of velocity space, the contributions to the integrals V_U and V_L in this region cancel for odd polynomials. Consequently there is no need to calculate that part of the sum.

The second point concerns the choice of points $\{v, \theta\}$, given s_l, at which to calculate the local Rosenbluth potentials g, given by (3.1.60), and h, given by (3.1.61). For a given orbit, it is a particular combination of these local potentials and their derivatives as expressed by (3.1.50) which determines the local Fokker–Planck coefficients shown in (3.1.59). Since it is a particular weighted sum of these local coefficients *along a given gyro-center trajectory* which corresponds to the bounce-averages indicated in (3.1.71) the natural choice is that given by (3.1.47): namely, choose $\{v, \theta\} = \{v_j, \theta(\theta_i, s_l)\}$. This corresponds to choosing local θ meshpoints which, traced back along an orbits, will pierce the midplane at a midplane meshpoint θ_i. This removes the need to interpolate during the bounce-average procedure and expedites the calculation.

3.2.3. Flux conservation at the trapped/passing boundary

Any consistent numerical scheme must satisfy condition (3.1.119), flux continuity, in the limit as the boundary layer becomes vanishingly thin. It remains to show the manner in which this is to be enforced numerically.

We will let mesh indices *itl* and *itu* prescribe the pitch-angles corresponding, respectively, to the co-moving and counter-moving legs of orbits on the trapped/passing boundary, "pinch" orbits: $\theta_{itu} = \pi - \theta_{itl}$. Now as the boundary layer about $\cos \theta_{itl} = -\cos \theta_{itu} = \mu_{0_T}$ becomes vanishingly thin, condition (3.1.119) must hold; this implies that the total density in the layer, i.e., the field line density in the sense of (3.1.54), must vanish identically in time. However, since the finite-difference mesh contains a boundary layer of finite thickness the density in this boundary layer will neither vanish nor remain stationary. In any event, a proper treatment of the boundary layer must account for the logarithmic singularity in τ_B there. This subject is treated in detail in Appendix

3B. Here we simply relate the results of that analysis to its implementation in the numerical scheme.

Generally speaking, the procedure used to evolve the distribution function in the boundary layer represents the processes within the finite width layer as suitable averages over the regions adjacent to the layer. To clearly specify how this is to be done we first need to show what is meant by "volume element" in the boundary layer. Then "suitable average" can be described as a specific algorithm motivated by considerations in Section 3.1.3 and Appendix 3B.

3.2.3.1. *Boundary layer: volume element*

As was shown through (3.1.54) and discussed in more detail in Appendix 3A, the conserved quantity in the bounce-averaged system is the field line density. We write

$$N = \int_0^{s_{mx}} \frac{ds}{\psi} n(s) = \int d^3v_0 \, v_{\|0} \tau_B(v_0, \theta_0) \mathscr{F}_0(v_0, \theta_0)$$

$$\cong \int_{\mathscr{R}} d^3v_0 \, v_{\|0} \tau_B(v_0, \theta_0) \mathscr{F}_0(v_0, \theta_0) + N_{bl}. \tag{3.2.7}$$

Here $\mathscr{R}$ denotes an integral whose domain is restricted away from the boundary layer defining the transition from passing to trapped particle orbits, and N_{bl}, the boundary layer line density, is given by

$$N_{bl} = 2\pi \int_{bl} dv_0 \, d\theta_0 \, v_0^2 \sin \theta_0 \lambda \mathscr{F}_0 = 4\pi \int_0^\infty dv_0 \, v_0^3 I_{bl}, \tag{3.2.8}$$

where I_{bl} is derived in Appendix 3B and $\lambda = v_{\|0} \tau_B$. The apparent extra factor 2 in (3.2.8) recognizes that the boundary layer has upper and lower halves, only one of which is represented explicitly by I_{bl}. Taking the time derivative of (3.2.8) and restructuring using the notation of Appendix 3B, (3B.2)–(3B.4), in particular

$$\mathscr{F}_{0_T} = \mathscr{F}_0(v_0, \theta_{itl}) = \mathscr{F}_0(v_0, \theta_{itu}),$$

there results

$$\frac{\partial N_{bl}}{\partial t} = 4\pi \int_0^\infty dv_0 \, v_0^3 \frac{I_{bl}}{\mathscr{F}_{0_T}} \frac{\partial \mathscr{F}_{0_T}}{\partial t}. \tag{3.2.9}$$

Defining the quantity

$$\mathscr{L}_{bl} = \frac{2 I_{bl} v_0}{\mathscr{F}_{0_T}}, \tag{3.2.10}$$

(3.2.9) may be re-expressed as

$$\frac{\partial N_{bl}}{\partial t} = 2\pi \int_0^\infty dv_0 \, v_0^2 \mathscr{L}_{bl} \frac{\partial \mathscr{F}_{0_T}}{\partial t} \tag{3.2.11}$$

and $\mathscr{L}_{bl}$ can be seen to play the role of $v_{\|0} \tau_B \sin \theta_0 \, d\theta_0 \equiv \lambda \sin \theta_0 \, d\theta_0$.

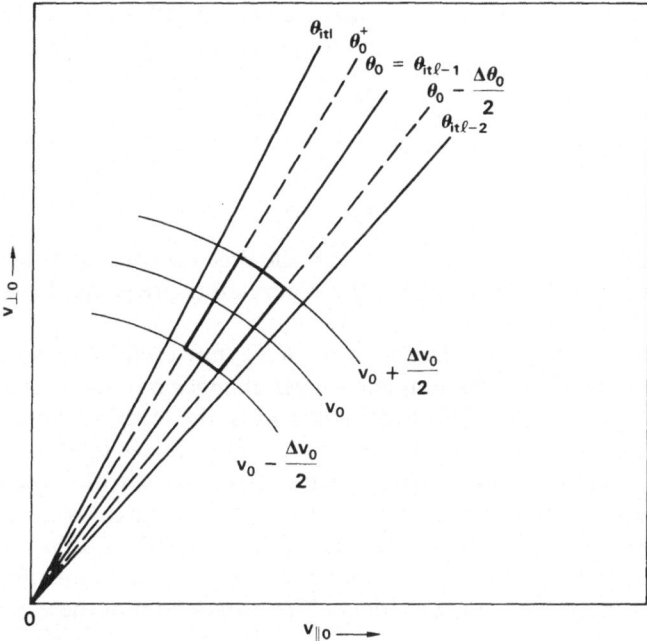

Figure 3.13. Boundary layer geometry. One of four volume elements adjacent to the boundry separating trapped and passing orbits is detailed. Orbit ensemble averages of fluxes in these volume elements are employed to define fluxes in and into the boundary layer (see Section 3.2.3 and Appendix 3B).

3.2.3.2. Boundary layer: averaging

In Appendix 3B, the edges of the boundary layer are defined by $\mu_{0_T}^{\pm}$. For purposes of the following we establish the following equivalences:

$$\theta_{itl}^{\mp} = \arccos \mu_{0_T}^{\pm} \qquad (3.2.12)$$

are the *lower* pitch-angle borders of the boundary layer, θ_{itl}^{-} lying in the co-passing orbit region and θ_{itl}^{+} lying on the co-moving side of the trapped orbit region (see Fig. 3.13). Also, $\theta_{itu}^{\mp}$ are the *upper* pitch-angle borders of the boundary layer defined analogously. The upper and lower borders (in velocity) of a volume element centered at v_0 are located at

$$v_0^{\pm} = v_0 \pm \tfrac{1}{2}\Delta v_0. \qquad (3.2.13)$$

Let the element in the co-passing region adjacent to the boundary layer be centered at θ_0. The pitch-angle borders of this element (see Fig. 3.13) are

$$\theta_0^{\pm} = \theta_0 \pm \tfrac{1}{2}\Delta\theta_0, \qquad (3.2.14)$$

where, of course, $\theta_0^{+} \equiv \theta_{itl}^{-}$.

The bounce-averaged Fokker–Planck equation (3.1.55) is valid at the point (v_0, θ_0) and we can integrate (3.1.55) across the element centered there to obtain

$$\frac{\partial N}{\partial t}(v_0, \theta_0) = 2\pi\gamma[\sin\theta_0\Delta\theta_0(\mathcal{G}_0(v_0^+, \theta_0) - \mathcal{G}_0(v_0^-, \theta_0))$$

$$+ \Delta v_0(\mathcal{H}_0(v_0, \theta_{itl}^-) - \mathcal{H}_0(v_0, \theta_0^-))].$$

This is simply a statement of the fact that the time rate of change of the number of particles in a volume element is the sum of all the relevant fluxes across the boundaries of the element.

The equation governing the evolution of $\mathcal{F}_{0_T}$ must reflect the constraint that the total line density in the boundary layer changes by exactly the sum of the θ_0-fluxes of all the volume elements adjacent to the layer. In deference to this, v_0-fluxes accorded by the "suitable average" of v_0-fluxes in elements adjacent to the layer can radistribute particles with respect to v_0 within the layer.[1]

The time rate of change of the total number of particles in a differential boundary volume element due to the θ_0-flux from the co-passing region is $-2\pi\gamma\mathcal{H}_0(v_0, \theta_{itl}^-)\Delta v_0$. There are similar expressions for contributions from counter-passing and trapped particles. The proper sum of these contributions has the form

$$2\pi\gamma\Delta v_0(-\mathcal{H}_0(v_0, \theta_{itl}^-) + 2\mathcal{H}_0(v_0, \theta_{itl}^+) + \mathcal{H}_0(v_0, \theta_{itu}^+)).$$

With regard to the v_0-fluxes, the relevant flux difference can be expressed as

$$2\pi\gamma(\langle\mathcal{G}_0\rangle_{bl}(v_0^+) - \langle\mathcal{G}_0\rangle_{bl}(v_0^-)),$$

where it remains to define the averaged v_0-flux, $\langle\mathcal{G}_0\rangle_{bl}$. While $\mathcal{F}_0$ and $\partial\mathcal{F}_0/\partial v_0$ are continuous across the boundary, $\partial\mathcal{F}_0/\partial\theta_0$ need not be. Furthermore, there may be discontinuities in the coefficients B_0 and C_0. We may extend the definition of λ to include the boundary layer region through

$$\lambda(\theta_{itl}) = \mathcal{L}_{bl}/2\sin\theta_{itl}\Delta\theta_{itl},$$

where $\Delta\theta_{itl}$ is the θ-width of the boundary region. Using this notation we can define

$$A_0(\theta_{itl}) = A_0(\theta_{itu}) \equiv \langle A_0\rangle_{bl}$$

$$\equiv \tfrac{1}{4}\lambda(\theta_{itl})\left(\frac{A_0}{\lambda}(\theta_{itl}^-) + \frac{A_0}{\lambda}(\theta_{itl}^+) + \frac{A_0}{\lambda}(\theta_{itu}^-) + \frac{A_0}{\lambda}(\theta_{itu}^+)\right) \qquad (3.2.15)$$

at velocity v_0. The expression for $\langle B_0\rangle_{bl}$ is similar. Terms of $\langle\mathcal{G}_0\rangle_{bl}$ in $\mathcal{F}_0$ and its v_0-derivative are thus unambiguously defined. Next, allowing for a discontinuity in $\partial\mathcal{F}_0/\partial\theta_0$ at θ_{itl} (and θ_{itu}), the associated average is

---

[1] Since $\mathcal{G}_0$, in general, may be discontinous (ideal slip) across the layer, this averaging introduces a certain traction or mixing within the layer. However, since we are not concerned with the internal structure of the layer per se, this is inconsequential to the calculation as a whole.

$$C_0(\theta_{itl})\frac{\partial \mathscr{F}_0}{\partial \theta_0}(\theta_{itl}) = C_0(\theta_{itu})\frac{\partial \mathscr{F}_0}{\partial \theta_0}(\theta_{itu}) \equiv \left\langle C_0 \frac{\partial \mathscr{F}_0}{\partial \theta_0} \right\rangle_{bl}$$

$$\equiv \tfrac{1}{4}\lambda(\theta_{itl})\left(\frac{C_0}{\lambda}(\theta_{itl}^-)\frac{\partial \mathscr{F}_0}{\partial \theta_0}(\theta_{itl}^-) + \frac{C_0}{\lambda}(\theta_{itl}^+)\frac{\partial \mathscr{F}_0}{\partial \theta_0}(\theta_{itl}^+)\right. \tag{3.2.16}$$

$$\left. + \frac{C_0}{\lambda}(\theta_{itu}^-)\frac{\partial \mathscr{F}_0}{\partial \theta_0}(\theta_{itu}^-) + \frac{C_0}{\lambda}(\theta_{itu}^+)\frac{\partial \mathscr{F}_0}{\partial \theta_0}(\theta_{itu}^+)\right).$$

Using (3.2.15) and (3.2.16) we can now represent $\langle \mathscr{G}_0 \rangle_{bl}$ as

$$\mathscr{G}_0(v_0, \theta_{itl}) = \mathscr{G}_0(v_0, \theta_{itu}) \equiv \langle \mathscr{G}_0 \rangle_{bl}$$

$$= \langle A_0 \rangle_{bl}\mathscr{F}_{0T} + \langle B_0 \rangle_{bl}\frac{\partial \mathscr{F}_{0T}}{\partial v_0} + \left\langle C_0 \frac{\partial \mathscr{F}_{0T}}{\partial \theta_0} \right\rangle_{bl}. \tag{3.2.17}$$

Multiplying these fluxes by the appropriate finite differentials, and combining, produces a conservation law for the boundary layer which we write as

$$\frac{\partial N}{\partial t}(v_0) = 2\pi\gamma[\Delta v_0(-\mathscr{H}_0(v_0, \theta_{itl}^-) + 2\mathscr{H}_0(v_0, \theta_{itl}^+) + \mathscr{H}_0(v_0, \theta_{itu}^+))$$

$$+ (\sin\theta_{itl}\Delta\theta_{itl} + \sin\theta_{itu}\Delta\theta_{itu})(\langle \mathscr{G}_0 \rangle_{bl}(v_0^+) - \langle \mathscr{G}_0 \rangle_{bl}(v_0^-))]. \tag{3.2.18}$$

Finally, dividing both sides of (3.2.18) by the boundary layer volume element from (3.2.11), and passing to the limit as $\Delta v_0 \to 0$, there results the evolution equation for $\mathscr{F}_{0T}$

$$\frac{\partial \mathscr{F}_{0T}}{\partial t}(v_0, \theta_{itl}) = \frac{\gamma}{\mathscr{L}_{bl}v_0^2}\left[(-\mathscr{H}_0(v_0, \theta_{itl}^-) + 2\mathscr{H}_0(v_0, \theta_{itl}^+) + \mathscr{H}_0(v_0, \theta_{itu}^+))\right.$$

$$\left. + (\sin\theta_{itl}\Delta\theta_{itl} + \sin\theta_{itu}\Delta\theta_{itu})\frac{\partial}{\partial v_0}\langle \mathscr{G}_0 \rangle_{bl}(v_0)\right]. \tag{3.2.19}$$

An examination of the limit of (3.2.19) as $\Delta\theta_{itl} = \Delta\theta_{itu} = \theta_{itl}^+ - \theta_{itl}^- \to 0$ shows it reduces to the jump conditon (3.1.119) as is required for consistency.

3.2.4. Implicit time advancement: operator splitting

The Fokker–Planck equation (3.1.49) with source and loss terms included may be written in the form (cf. (3.1.55))

$$\frac{\partial \lambda\mathscr{F}_0}{\partial t} = \gamma\left(\frac{1}{v_0^2}\frac{\partial \mathscr{G}_0}{\partial v_0} + \frac{1}{v_0^2\sin\theta_0}\frac{\partial \mathscr{H}_0}{\partial \theta_0}\right) + \lambda\hat{K}\mathscr{F}_0 + \lambda S, \tag{3.2.20}$$

where $\mathscr{G}_0$ and $\mathscr{H}_0$ incorporate collisional, quasilinear, and ohmic terms. Here S is a generalized source of the type described in Section 3.1.4 and $\hat{K}$ is a Krook operator representing various loss terms as described in Section 3.1.5.

For notational simplicity we suppress the subscript on variables $\mathscr{F}_0$, v_0, and θ_0 for purposes of writing finite-difference equations. It should be realized, however, that all the differencing is done in the midplane system.

As before we construct pitch-angle and velocity meshes corresponding to

$$\{\theta_i, i = 1, I\}; \qquad \{v_j, \; j = 1, J\}.$$

While the mesh is required to be symmetric about $\frac{1}{2}\pi$, there is no mesh point at $\frac{1}{2}\pi$. We will use the convention in the following that θ_M denotes the largest θ meshpoint less than $\frac{1}{2}\pi$.

Velocity differences come in the two varieties

$$\Delta v_j = \tfrac{1}{2}(v_{j+1} - v_{j-1}) \quad \text{and} \quad \Delta v_{j+1/2} = (v_{j+1} - v_j),$$

quantities determined at half meshpoints obey

$$B_{j+1/2} = \tfrac{1}{2}(B_j + B_{j+1}),$$

and we will represent the distribution function on the mesh by the notation

$$\mathscr{F}_{i,j} = \mathscr{F}(v_j, \theta_i).$$

Similarly, θ differences are written

$$\Delta\theta_i = \tfrac{1}{2}(\theta_{i+1} - \theta_{i-1}) \quad \text{and} \quad \Delta\theta_{i+1/2} = (\theta_{i+1} - \theta_i).$$

The boundary regions about θ_{itl} and θ_{itu} are chosen as thin slivers typically 5% of the θ mesh spacing elsewhere. This aids the accuracy of the technique used to treat the boundary layer as described in Appendix 3B and Section 3.2.3.

An operator splitting scheme is used to time advance the kinetic equation (3.2.20). While the technique is well known (Richtmyer and Morton, 1967), its application in this case is somewhat nonstandard owing to the nonlocal nature of the differential operator due to the orbit topology (cf. Section 3.1.3). In this connection, certain features of the bounce-averaged system are unique and warrant description in their own right. Therefore, the procedure will be described in some detail here with particular emphasis on the manner in which density conservation is assured.

The conserved quantity, the numerical approximation to the field line density N defined by (3.2.7), is given by

$$N = \sum_{i=1}^{I} \sum_{j=1}^{J} a_i b_j \lambda_i \mathscr{F}_{i,j}, \qquad (3.2.21)$$

where

$$a_i = \begin{cases} 2\pi \sin\theta_i \Delta\theta_i & \text{for } i \notin \{1, I\}, \\ \pi\theta_{3/2}^2 & \text{for } i = 1, \\ \pi(\pi - \theta_{I-1/2})^2 & \text{for } i = I, \end{cases}$$

$$b_j = \begin{cases} v_j^2 \Delta v_j & \text{for } j \notin \{1, J\}, \\ \tfrac{1}{3}v_{3/2}^2 & \text{for } j = 1, \end{cases}$$

and where we require that $\mathscr{F}$ vanish at the upper velocity extent of the mesh, $\mathscr{F}_{i,J} = 0$. Further, we have defined

$$\lambda_i = \begin{cases} v_j \cos\theta_i \tau_B(v_j, \theta_i) & \text{for } i \notin \{itl, itu\}, \\ \mathscr{L}_{bl}/2 \sin\theta_{itl}\Delta\theta_{itl} & \text{for } i \in \{itu, itl\}. \end{cases}$$

The numerical equations employed for each half of the operator splitting scheme are structured to conserve particle density exactly in the absence of particle sources and sinks. The equations describing the first half of the procedure, the split in the velocity v direction are

$$\lambda_i \mathscr{F}_{i,j}^{n+1/2} = \begin{cases} \lambda_i \mathscr{F}_{i,j}^n + \dfrac{\gamma \Delta t}{v_j^2 \Delta v_j}(\mathscr{G}_{i,j+1/2}^{n+1/2} - \mathscr{G}_{i,j-1/2}^{n+1/2}) \\[2mm] \qquad + \dfrac{\Delta t}{2}\lambda_i \hat{K}_{i,j}^n \mathscr{F}_{i,j}^{n+1/2} + \dfrac{\Delta t}{2}\lambda_i S_{i,j}^n \end{cases} \text{ for } \begin{cases} \forall i \\ 1 < j < J, \end{cases}$$

for interior meshpoints, and

$$\lambda_i \mathscr{F}_{i,j}^{n+1/2} = \begin{cases} 0 & \text{ for } \begin{cases} \forall i \\ j = J, \end{cases} \\[4mm] \lambda_i \mathscr{F}_{i,1}^n + \dfrac{3\gamma \Delta t}{v_{3/2}^3}\mathscr{G}_{i,3/2}^{n+1/2} \\[2mm] \qquad + \dfrac{\Delta t}{2}\lambda_i \hat{K}_{i,1}^n \mathscr{F}_{i,1}^n + \dfrac{\Delta t}{2}\lambda_i S_{i,1}^n & \text{ for } \begin{cases} \forall i \\ j = 1, \end{cases} \end{cases} \quad (3.2.22)$$

for boundary points. Furthermore, the symmetry constraint in the boundary region forces

$$\mathscr{F}_{itu,j}^{n+1/2} = \mathscr{F}_{itl,j}^{n+1/2} \qquad \forall j. \tag{3.2.23}$$

In the preceding, the v-flux $\mathscr{G}$ is given in terms of the coefficients and divided differences as

$$\mathscr{G}_{i,j+1/2}^{n+1/2} = \begin{cases} (A^n \mathscr{F}^{n+1/2})_{i,j+1/2} + B_{i,j+1/2}^n \dfrac{\mathscr{F}_{i,j+1}^{n+1/2} - \mathscr{F}_{i,j}^{n+1/2}}{\Delta v_{j+1/2}} & \text{for } i \notin \{itl, itu\}, \\[4mm] \quad + C_{i,j+1/2}^n \dfrac{\mathscr{F}_{i+1,j+1/2}^n - \mathscr{F}_{i-1,j+1/2}^n}{2\Delta\theta_i} \\[4mm] (A^n \mathscr{F}^{n+1/2})_{itl,j+1/2} + B_{itl,j+1/2}^n \dfrac{\mathscr{F}_{itl,j+1}^{n+1/2} - \mathscr{F}_{itl,j}^{n+1/2}}{\Delta v_{j+1/2}} \\[4mm] \quad + \tfrac{1}{4}\lambda_{itl}\left\{\left(\dfrac{C}{\lambda}\right)_{itl-1,j}^n \dfrac{\mathscr{F}_{itl,j}^n - \mathscr{F}_{itl-1,j}^n}{\Delta\theta_{itl-1/2}}\right. \\[4mm] \quad + \left(\dfrac{C}{\lambda}\right)_{itl+1,j}^n \dfrac{\mathscr{F}_{itl+1,j}^n - \mathscr{F}_{itl,j}^n}{\Delta\theta_{itl+1/2}} & \text{for } i \in \{itl, itu\}, \\[4mm] \quad + \left(\dfrac{C}{\lambda}\right)_{itu-1,j}^n \dfrac{\mathscr{F}_{itu,j}^n - \mathscr{F}_{itu-1,j}^n}{\Delta\theta_{itu-1/2}} \\[4mm] \quad + \left. \left(\dfrac{C}{\lambda}\right)_{itu+1,j}^n \dfrac{\mathscr{F}_{itu+1,j}^n - \mathscr{F}_{itu,j}^n}{\Delta\theta_{itu+1/2}}\right\} \end{cases} \quad (3.2.24)$$

where the averages have been performed as indicated by (3.2.15), (3.2.16), and (3.2.17) over the meshpoints nearest to θ_{itl} and θ_{itu}.

It can be shown that $C_{i,j+1/2} = 0$ for $i \in \{1, I\}$. Also, the superscripts $n + \frac{1}{2}$ and $n + 1$ are used to indicate time-dependent quantities after the first (v) and second (θ) splits, respectively. The cross derivatives are explicit and $\mathscr{F}$ is defined uniquely at the origin only after the second split.

That (3.2.22) conserves density can be verified by forming the numerical derivative of (3.2.21):

$$N^{n+1/2} - N^n = \sum_{i=1}^{I} \sum_{j=1}^{J} a_i b_j \lambda_i (\mathscr{F}_{i,j}^{n+1/2} - \mathscr{F}_{i,j}^n). \tag{3.2.25}$$

Eliminating the source and sink terms in (3.2.22) and substituting into (3.2.25) gives

$$N^{n+1/2} = N^n.$$

The second half of the operator splitting scheme, now in the θ-direction, is represented by the following: the interior points are advanced by

$$\lambda_i \mathscr{F}_{i,j}^{n+1} = \begin{cases} \lambda_i \mathscr{F}_{i,j}^{n+1/2} + \dfrac{\gamma \Delta t}{v_j^2 \sin \theta_i \Delta \theta_i} (\mathscr{H}_{i+1/2,j}^{n+1} - \mathscr{H}_{i-1/2,j}^{n+1}) \\[3mm] \qquad + \dfrac{\Delta t}{2} \lambda_i (\hat{K}_{i,j}^n \mathscr{F}_{i,j}^{n+1} + S_{i,j}^n) \end{cases} \quad \text{for} \begin{cases} i \notin \{itl, itu\}, \\ i \notin \{1, m, I\}, \\ \forall j, \end{cases}$$

while the boundary points are advanced by

$$\lambda_i \mathscr{F}_{i,j}^{n+1} = \begin{cases} \lambda_1 \mathscr{F}_{1,j}^{n+1/2} + \dfrac{2\gamma \Delta t}{v_j^2 \theta_{3/2}^2} \mathscr{H}_{3/2,j}^{n+1} \\[2mm] \qquad + \dfrac{\Delta t}{2} \lambda_1 (\hat{K}_{1,j}^n \mathscr{F}_{1,j}^{n+1} + S_{1,j}^n) & \text{for} \begin{cases} i = 1, \\ \forall j, \end{cases} \\[7mm] \lambda_m \mathscr{F}_{m,j}^{n+1/2} - \dfrac{\gamma \Delta t}{v_j^2 \sin \theta_m \Delta \theta_m} \mathscr{H}_{m-1/2,j}^{n+1} \\[2mm] \qquad + \dfrac{\Delta t}{2} \lambda_m (\hat{K}_{m,j}^n \mathscr{F}_{m,j}^{n+1} + S_{m,j}^n) & \text{for} \begin{cases} i = m, \\ \forall j, \end{cases} \\[7mm] \lambda_I \mathscr{F}_{I,j}^{n+1/2} - \dfrac{2\gamma \Delta t}{v_j^2 (\pi - \theta_{I-1/2})^2} \mathscr{H}_{I-1/2,j}^{n+1} \\[2mm] \qquad + \dfrac{\Delta t}{2} \lambda_I (\hat{K}_{I,j}^n \mathscr{F}_{I,j}^{n+1} + S_{I,j}^n) & \text{for} \begin{cases} i = I, \\ \forall j, \end{cases} \\[7mm] \lambda_{itl} \mathscr{F}_{itl,j}^{n+1/2} - \dfrac{\gamma \Delta t}{2v_j^2 \sin \theta_{itl} \Delta \theta_{itl}} \\[2mm] \qquad \times (\mathscr{H}_{itl-1/2,j}^{n+1} - 2\mathscr{H}_{itl+1/2,j}^{n+1} - \mathscr{H}_{itu+1/2,j}^{n+1}) & \text{for} \begin{cases} i = itl, \\ \forall j, \end{cases} \\[2mm] \qquad + \dfrac{\Delta t}{2} \lambda_{itl} (\hat{K}_{itl,j}^n \mathscr{F}_{itl,j}^{n+1} + S_{itl,j}^n) \end{cases} \tag{3.2.26}$$

The value of $\mathscr{F}$ at the origin will be discussed presently. In the algorithm (3.2.26) we have defined

$$\mathscr{H}_{i+1/2,j}^{n+1} = (D^n \mathscr{F}^{n+1})_{i+1/2,j} + E_{i+1/2,j}^n \frac{\mathscr{F}_{i+1/2,j+1}^{n+1/2} - \mathscr{F}_{i+1/2,j-1}^{n+1/2}}{2\Delta v_j}$$
$$+ F_{i+1/2,j}^n \frac{\mathscr{F}_{i+1,j}^{n+1} - \mathscr{F}_{i,j}^{n+1}}{\Delta\theta_{i+1/2}}. \tag{3.2.27}$$

Density is conserved identically during the second half of the splitting algorithm as in the first half. This may be shown by a calculation similar to that indicated by (3.2.25).

There remains the matter of a unique value for $\mathscr{F}$ at the origin, $v_0 = 0$. This is accomplished by requiring that

$$\mathscr{F}_{i,1}^{n+1} = \frac{\sum_{i=1}^{I} a_i \lambda_i \mathscr{F}_{i,1}^{n+1/2}}{\sum_{i=1}^{I} a_i \lambda_i} \qquad \forall i. \tag{3.2.28}$$

The algorithm for solving the two sets of equations (3.2.22) and (3.2.26) representing three-coupled tridiagonal systems, is standard. However, the coupling of the three systems, corresponding to the co- and counter-passing regions and the trapped region of velocity space, is somewhat uncommon and is presented here in the interest of completeness.

Equations (3.2.22) may be cast in the following canonical form:

$$-\alpha_{i,j}^n \mathscr{F}_{i,j+1}^{n+1/2} + \beta_{i,j}^n \mathscr{F}_{i,j}^{n+1/2} - \gamma_{i,j}^n \mathscr{F}_{i,j-1}^{n+1/2} = \delta_{i,j}^n, \tag{3.2.29}$$

where α, β, γ, and δ are known at timestep n. The indices i, j range over all values on the mesh; vacuous terms such as $\gamma_{1,1}^n$ are ignored since $\mathscr{F}_{i,0}$ does not exist.

The standard procedure for solving (3.2.29), involving two sweeps through the mesh, can be found in Richtmyer and Morton (1967). Let

$$\mathscr{F}_{i,j}^{n+1/2} = \sigma_{i,j} \mathscr{F}_{i,j+1}^{n+1/2} + \tau_{i,j}, \qquad j = 1, \ldots, J-1, \tag{3.2.30}$$

then (3.2.29) yields

$$\sigma_{i,j} = (\beta_{i,j}^n - \gamma_{i,j}^n \sigma_{i,j-1})^{-1} \alpha_{i,j}^n \tag{3.2.31}$$

and

$$\tau_{i,j} = (\beta_{i,j}^n - \gamma_{i,j}^n \sigma_{i,j-1})^{-1} (\delta_{i,j}^n + \gamma_{i,j}^n \tau_{i,j-1}). \tag{3.2.32}$$

The quantities $\sigma_{i,1}$ and $\tau_{i,1}$, determined through the relevant boundary relation $(j = 1)$ in (3.2.22), allow initialization of the recurrence relation for $j > 1$. The condition for $(j = J)$ in (3.2.22) determines $\mathscr{F}_{i,J}$ thus initializing the back solve to evaluate $\mathscr{F}_{i,j}$ for $j < J$.

The velocity split thus involves only two sweeps (a forward recursion, then a back solve) over the index j for each value of index i. The pitch-angle split, however, involves six sweeps over the index i for each value of index j. Each sweep–solve pair corresponds to one of the three sectors of velocity space joined at the boundary layer; namely, $[0, \theta_{itl}]$, $[\frac{1}{2}\pi, \theta_{itl}]$, and $[\pi, \theta_{itu}]$. The first three boundary relations in (3.2.26) are used to initialize each of the three forward sweeps to determine the $\{\sigma\}$ and the $\{\tau\}$. Then the last condition of (3.2.26), which links the regions at the boundary layer, is used to initialize each respective back solve.

Schematically, the procedure is as follows: with the exception of the cases $i \in \{itl, itu\}$ the difference equations (3.2.26) are cast in the form

$$-\varepsilon_{i,j}^n \mathscr{F}_{i+1,j}^{n+1} + \mu_{i,j}^n \mathscr{F}_{i,j}^{n+1} - v_{i,j}^n \mathscr{F}_{i-1,j}^{n+1} = \varrho_{i,j}^n \tag{3.2.33}$$

(again ignoring vacuous terms such as v_{11}). At $i \in \{itl, itu\}$, (3.2.26) does not assume the form (3.2.33) since the boundary layer involves values of $\mathscr{F}$ at four points at the advanced time rather than three. Expressed in recursive form in the index i, (3.2.33) is treated in the same manner as (3.2.29) on each sector separately; viz., the forward recursion over $[\frac{1}{2}\pi, \theta_{itl}]$ is represented by

$$\mathscr{F}_{i,j}^{n+1} = \sigma_{i,j} \mathscr{F}_{i-1,j}^{n+1} + \tau_{i,j}; \quad i = itl + 1, \ldots, m.$$

The relevant $\{\sigma\}$ and $\{\tau\}$ are determined through recursive equations with $\sigma_{i,j}$, $\tau_{i,j}$ for $i \in \{1, m, I\}$ set by the boundary relations in (3.2.26).

One further piece of information is necessary to complete the algorithm. The value of $\mathscr{F}$ in the boundary layer given by the last relation in (3.2.26) is used to initialize the three back solve recursions as follows: Recasting that condition in the form

$$\mathscr{F}_{itl,j}^{n+1} = \varepsilon_{itl,j}^n \mathscr{F}_{itl+1,j}^{n+1} + \mu_{itl,j}^n \mathscr{F}_{itu+1,j}^{n+1} + v_{itl,j}^n \mathscr{F}_{itl-1,j}^{n+1} + \varrho_{itl,j}^n, \tag{3.2.34}$$

we substitute the recurrence formula for $\mathscr{F}_{itl,j}^{n+1}$ into (3.2.34) to obtain

$$\begin{aligned} \mathscr{F}_{itl,j}^{n+1} = {}& \varepsilon_{itl,j}^n(\sigma_{itl+1,j}\mathscr{F}_{itl,j}^{n+1} + \tau_{itl+1,j}) + \mu_{itl,j}^n(\sigma_{itu+1,j}\mathscr{F}_{itu,j}^{n+1} + \tau_{itu+1,j}) \\ & + v_{itl,j}^n(\sigma_{itl-1,j}\mathscr{F}_{itl,j}^{n+1} + \tau_{itl-1,j}) + \varrho_{itl,j}^n. \end{aligned} \tag{3.2.35}$$

This is equation for $\mathscr{F}_{itl,j}^{n+1}$ whose solution initialize the back solve recursions.

The drawbacks of operators splitting are primarily associated with limitations on the size of the timestep (in connection with stability constraints) since the scheme is not fully implicit.

Temporal evolution of the solution is the only practical way to the steady state, and in cases where the steady state is of principal interest, the algorithm is less than optimal. Kinetic equations of similar type have been solved with fully implicit differencing in a single step. The uniform field calculation can be case in the generic form

$$\mathsf{M}\mathbf{f}^{n+1} = \mathbf{g},$$

where $\mathbf{f}^{n+1}$ is a vector with $I \times J$ components corresponding to meshpoints ordered by row or column. The matrix M is nine-banded, a consequence of the nine-point difference algorithm, and can be inverted using the ICCG (incomplete Cholesky conjugate gradient) algorithm (see Section 2.4.1.8). The situation for bounce-averaged Fokker–Planck equations is complicated by the presence of the boundary layer where the orbit topology does not permit a simple nine-point difference algorithm. Incomplete LU decomposition with biconjugate gradient iteration packages that treat arbitrary sparsity patterns are indicated.

3.3. Applications

In this section the application of the code to two specific tokamak plasma problems is discussed. A visual introduction to the broad spectrum of plasma kinetic phenomena which the code can exhibit is first presented in the guise of a particular ICRF scenario. Details of the two examples are then developed, results are presented, and conclusions are drawn in each case.

The ICRF scenario we use for illustration follows the second harmonic RF excitation of a deuterium distribution in the presence of fixed Maxwellian background distributions. The steady state is achieved by a balance between resonant and collisional diffusion; energy is absorbed from the wave field by the deuterium distribution, then collisionally transferred to the fixed background deuterium and electron distributions.

The toroidal geometry in which the calculation takes place has the following dimensions: major radius $R_0 = 300$ cm, minor radius $a = 130$ cm, and flux surface minor radius $r = 65$ cm, yielding inverse aspect ratio $\varepsilon = 0.217$. The tritium background distribution is a 4.5 keV Maxwellian of density $\sim 10^{14}$ cc^{-1} at the midplane. The electrons preserve charge quasineutrality. The deuterium is originally thermal at 4.5 keV and 10^{14} cc^{-1} at the midplane. The RF wave field chosen employs $k_\parallel = 0.0255$ cm^{-1}, and $k_\perp = 0.67$ cm^{-1}. The electric field is adjusted dynamically so that the steady-state absorbed power is 1 W cc^{-1}.

Figure 3.14 shows $\langle\!\langle 1/\psi(s) \rangle\!\rangle$ as a function of $\cos\theta_0$, where $\psi(s)$ is defined throught (3.1.40). The minimum at the trapped/passing boundary is due to the fact that particles on such marginally trapped orbits spend almost all their time in the vicinity of $s = s_{mx}$ where ψ is maximal (the "throat"). Note however that while from (3B.5) $1/\psi(s_{mx}) = 0.71$, the minimum given by Fig. 3.14 is 0.720 because $\langle\!\langle 1/\psi \rangle\!\rangle$ has been effectively averaged over the ensemble of orbits in the boundary layer as described in Appendix 3B, not simply evaluated at the boundary.

Figure 3.15 shows the steady-state distribution function $\mathscr{F}$: the nonthermal tail is sustained by the RF; the average energy is 6.9 keV. Figure 3.16 gives a description of the dynamical steady-state from the perspective of the various velocity space fluxes which persist in that state. The figure is composed of three parts showing: (a) the RF flux alone; (b) the collisional flux alone; and (c) the combination of the two.

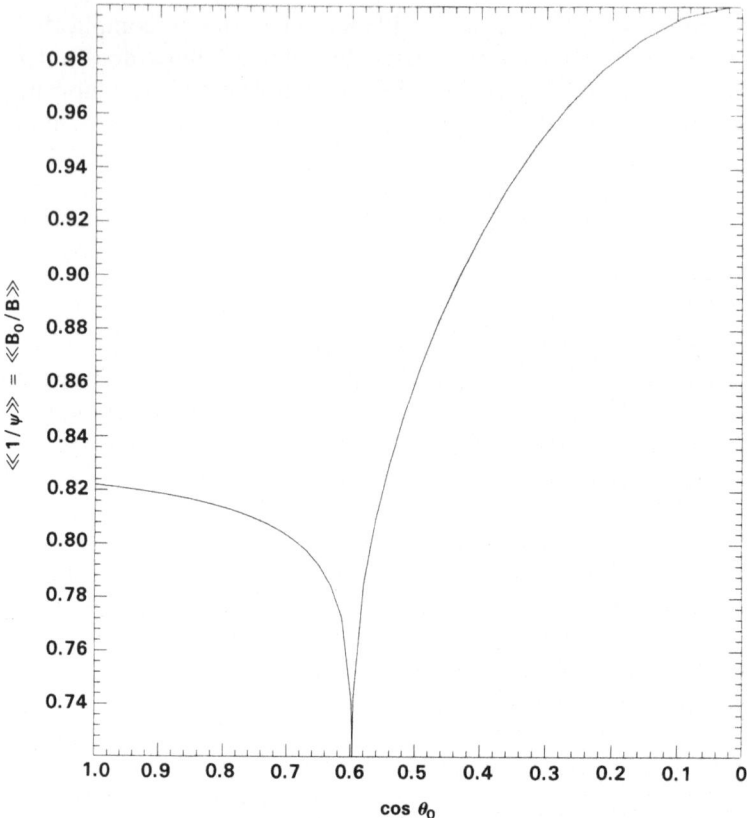

Figure 3.14. Bounce-average of $1/\psi$: $\langle\!\langle 1/\psi \rangle\!\rangle$. Represented as a function of $\mu = \cos\theta$ for a toroidal system (see (3.1.40)) of inverse aspect ratio $\varepsilon = 0.09$ (see also Section 3.3).

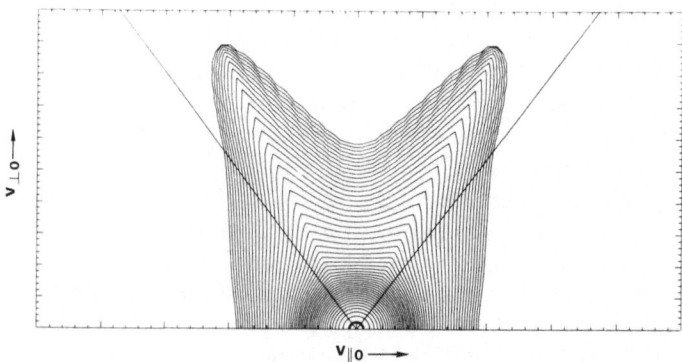

Figure 3.15. Contours of the steady-state ion distribution function $\mathscr{F}$ represented in midplane velocity space coordinates $(v_{\|0}, v_{\perp 0})$. Tail structure sustained by second harmonic RF excitation (see Section 3.3).

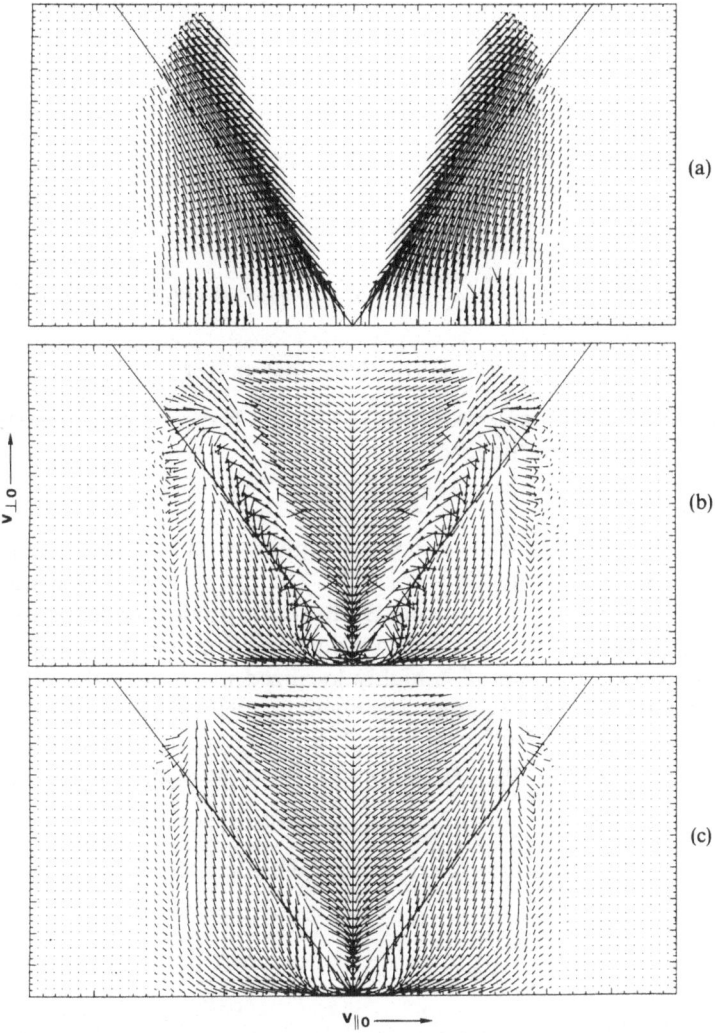

Figure 3.16. Steady-state velocity space fluxes. (a) RF flux; (b) RF + collisional flux; (c) collisional flux (see Section 3.3).

3.3.1. Neoclassical corrections to classical resistivity

The presence of trapped particles in a tokamak plasma has a marked effect on the electrical conductivity. Analysis by Rosenbluth *et al.* (1972) yields the expansion in inverse aspect ratio ε

$$\sigma = \sigma_{sp}(1.0 - 1.95\sqrt{\varepsilon} + O(\varepsilon)) \tag{3.3.1}$$

in the limit $\varepsilon \ll 1$. Here the classical "Spitzer" value (Spitzer, 1967) for the conductivity is

$$\sigma_{sp} = 2\frac{n_e e^2}{m_e}\tau_e \tag{3.3.2}$$

and the electron–ion momentum transfer time τ_e is

$$\tau_e = 3.44 \times 10^5 \frac{T_e^{3/2}}{n_e \lambda_{ei}}. \tag{3.3.3}$$

With further analysis Coppi and Sigmar (1973) have shown the $O(\varepsilon)$ correction in (3.3.1) to be given by

$$\sigma = \sigma_{sp}(1.0 - 1.95\sqrt{\varepsilon} + 0.95\varepsilon). \tag{3.3.4}$$

Connor et al. (1973) extended these estimates for finite ε. They express the distribution function in the form

$$f = f_s + \hat{f}, \tag{3.3.5}$$

where f_s is the distribution which would obtain in a uniform magnetic field calculation as described by Spitzer (1967). The *Spitzer–Härm* collision operator described in Spitzer and Härm (1953) can be used to describe the bulk of the passing particles correctly while a model collision operator is applied to $\hat{f}$ under the assumption that $\hat{f}$ is highly localized in pitch-angle. The analysis results in a value for the "average" conductivity, defined as $\bar{\sigma} = J_\phi/E_0$ where J_ϕ is the toroidal current density averaged over a magnetic surface and E_0 is the toroidal electric field at the magnetic axis. The value is given as

$$\bar{\sigma} = \sigma_{sp}I\left\{\frac{1 - 0.039(1 - I)}{1 + 0.471(1 - I)}\right\}, \tag{3.3.6}$$

where

$$I = \tfrac{3}{4}\langle B\rangle_{\theta_p}\int_0^{1/B_{mx}}\frac{\lambda\,d\lambda}{\langle\sqrt{1 - \lambda B/B}\rangle_{\theta_p}}. \tag{3.3.7}$$

Connor et al. (1973) make the point that as $\varepsilon \to 1$ the distribution of passing particles becomes localized so that an accurate result can be gleaned in that case by just applying the model operator directly to f rather than to $\hat{f}$. The associated limit is

$$\bar{\sigma} = 0.93I\sigma_{sp} \tag{3.3.8}$$

rather than the result of (3.3.6) which turns out to be

$$\bar{\sigma} = 0.71I\sigma_{sp}. \tag{3.3.9}$$

Those authors are confident in the accuracy of these results to within 20%.

The bounce-averaged Fokker–Planck code was used to compute resistivity enhancement $\bar{\eta}/\eta_{sp}$ due to toroidal effects over the full range of ε. The pro-

Figure 3.17. Ratio of resistivity (neoclassical value) to Spitzer resistivity (classical value), η/η_s as a function of inverse aspect ratio, ε: (a) solid line—code result; (b) upper broad dashed line—determined by (3.3.4); (c) fine dashed line—determined by the analysis of Connor et al. (1973): Dashed line beginning at $\varepsilon = 0.65$ is the Connor et al. (1973) result for the asymptotic case $\varepsilon \to 1$ (see Section 3.3.1).

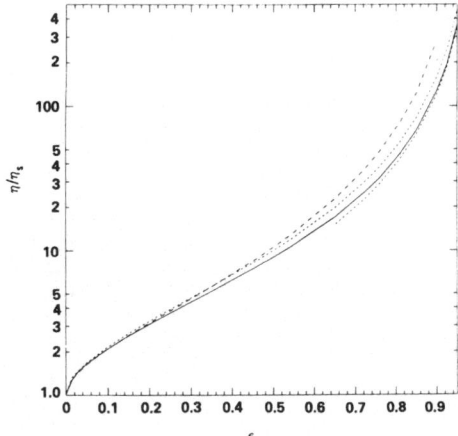

cedure employed was to choose an initial electron Maxwellian distribution and a fixed Maxwellian ion background distribution. The ions serve as a heat bath to allow the electrons to achieve a steady state. A small parallel electric field is applied, $E_0 \sim 0.03E_D$, where E_D is the Dreicer electric field (Dreicer, 1959)

$$E_D = \frac{m_e v_e v_e}{2e},$$

v_e is the electron thermal velocity, and v_e is the collision frequency

$$v_e = \frac{4\pi e^4 n_e \ln \Lambda_{ei}}{m_e^2 v_e^3}.$$

This choice of E_0 guarantees an exponentially small population of runaway electrons (electric force exceeds drag). Figure 3.17 shows the resistivity enhancement as a function of ε as determined by the code (solid), as determined by (3.3.4) (upper dashed), and as determined by the Connor et al. (1973) analysis (3.3.6) and (3.3.8) (fine dashed). Note the universal agreement at the $\varepsilon \to 0$ limit. The dashed line which starts as $\varepsilon \sim 0.65$ represents the Connor result for the asymptotic case $\varepsilon \to 1$. Agreement with the asymptotic limits is better than 3% in all cases in which agreement is expected.

The two Figs. 3.18 and 3.19 illustrate the steady-state ohmically driven distribution for the case $\varepsilon = 0.09$. Figure 3.18 shows: (a) the electric field flux; (b) the collisional flux; and (c) the combined flux. Note the watershed at large positive $v_{\parallel}$ in the collisional flux. Collisional diffusion at this point is entirely pitch-angle diffusion. As the electric field is increased, a larger proportion of electrons are found on runaway orbits. For $E_0 > 0.3E_D$ even an approximate steady-state becomes unachievable due to this effect in the absence of addi-

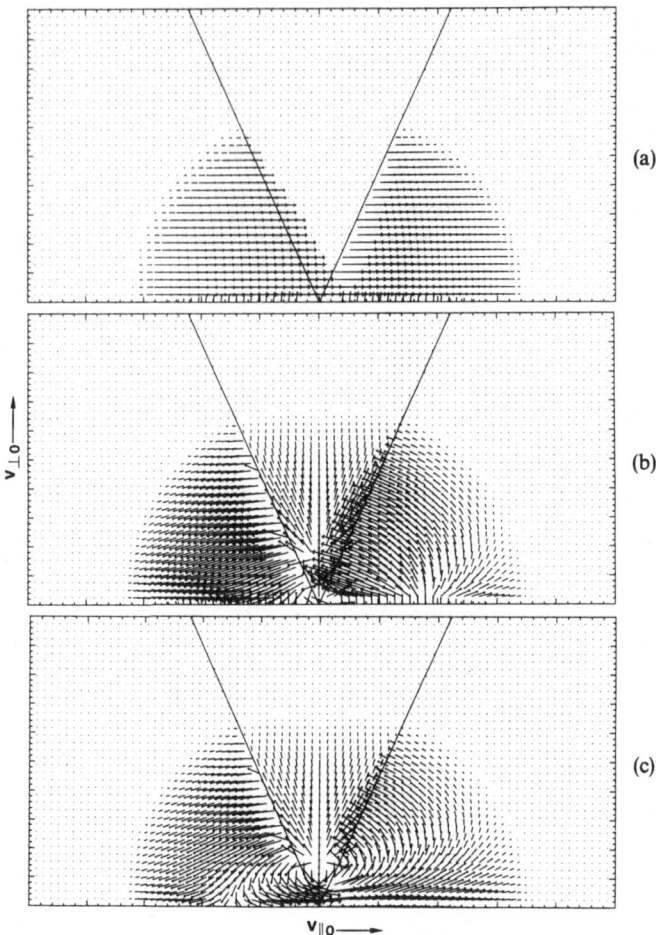

Figure 3.18. Steady-state velocity space fluxes. Fluxes at steady state for plasma resistivity calculations, $\varepsilon = 0.09$. (a) The parallel electric field or ohmic flux; (b) the collisional flux; (c) the sum of ohmic and collisional fluxes (see Section 3.3.1).

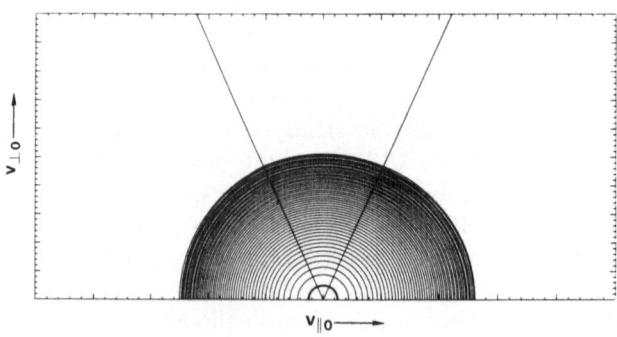

Figure 3.19. Contours of the electron distribution function $\mathscr{F}_e$ at near steady-state conditions for $\varepsilon = 0.09$.

Figure 3.20. $\mathscr{F}_\parallel = \int dv_\perp v_\perp \mathscr{F}$ versus $v_\parallel$. The (parallel velocity) distribution describes electrons in a large parallel electric field ($E_\parallel = 0.07$ V/cm^{-1}) after evolving from a Maxwellian for a period of $t \sim 50\tau_{ei}$. Note the positive slope at small $v_\parallel$ (see Section 3.3.1).

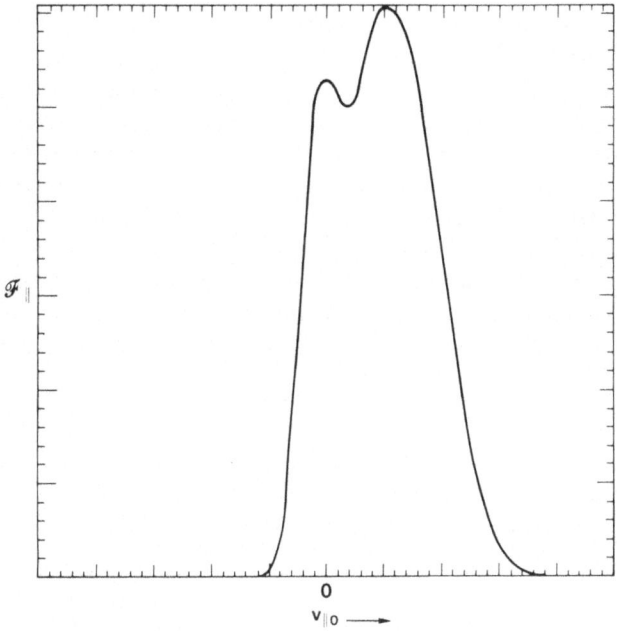

Figure 3.21. (a) Contours of distribution $\mathscr{F}$ for the same case as Fig. 3.20 at $t \sim 50\tau_{ei}$. (b) Contours of $\Delta t \partial \mathscr{F}/\partial t$ for the same case.

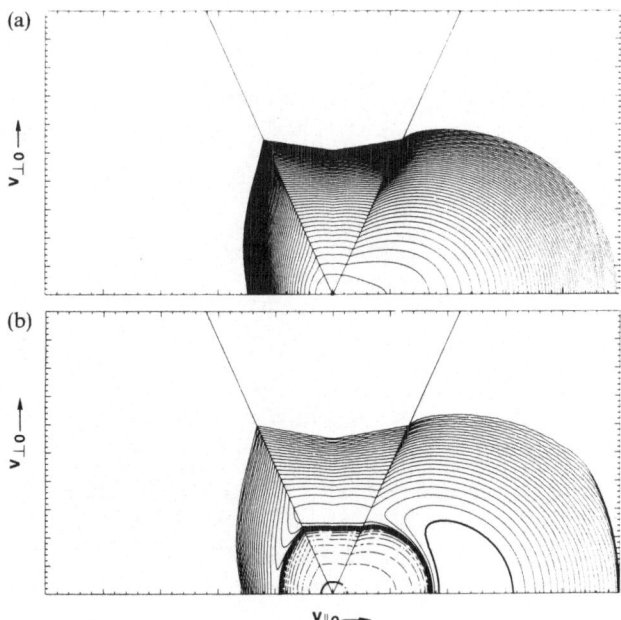

tional mitigating considerations. In fact, for these large electric fields, the concept of E_D becomes somewhat nebulous since it is defined in terms of the steady-state. Figure 3.19 shows contours of $\mathscr{F}$; the distribution looks nearly Maxwellian due to the minimal applied field ($E_\parallel = 0.003$ V cm^{-1}. When the electric field is increased, the parallel distribution ($\mathscr{F}$ integrated over $v_\perp$) exhibits a range of positive slope. The feature, characteristic of the "slideaway" regime described by Hui *et al.* (1977), can be recognized in Fig. 3.20 ($E_\parallel = 0.07$ V cm^{-1}, $\varepsilon = 0.09$). Figure 3.21 shows: (a) the distribution $\mathscr{F}$; and (b) $\partial \mathscr{F}/\partial t$ after a time equivalent to about $50\tau_e$. Lastly, Fig. 3.22 shows the relevant fluxes; the electric field flux (a) dominates the collisional flux (b) for most values of $v_\parallel > 0$ as evidenced by its similarity to the combined flux (c).

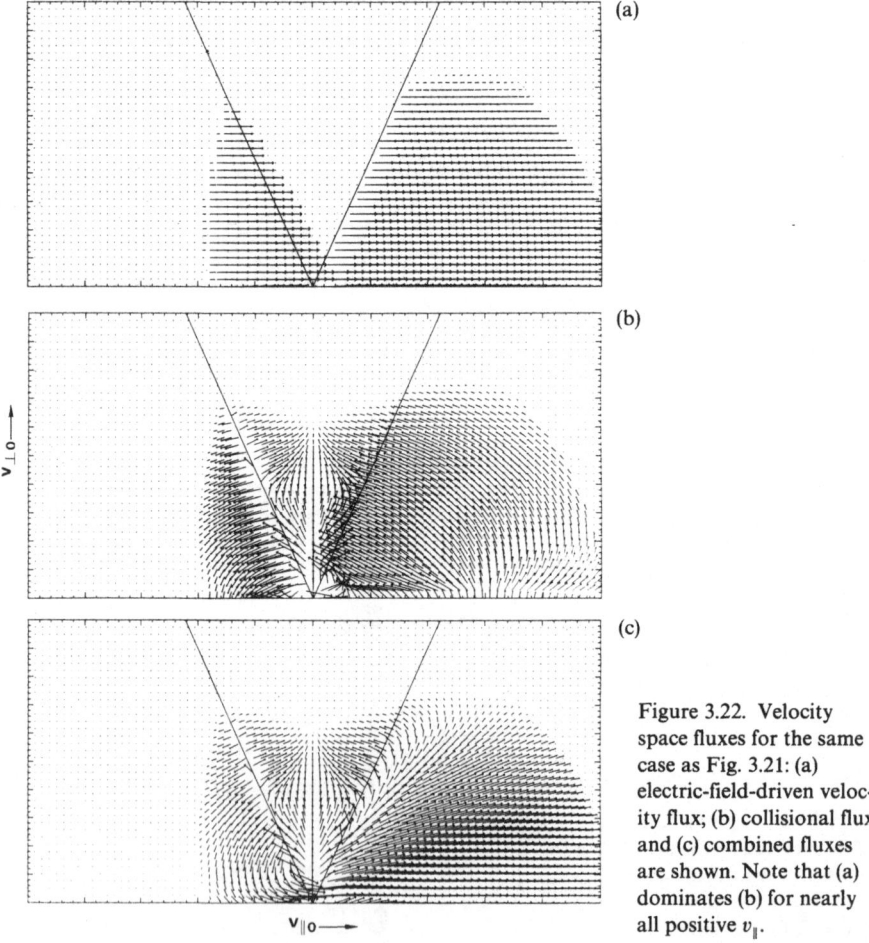

(a)

(b)

$v_\perp 0 \longrightarrow$

(c)

Figure 3.22. Velocity space fluxes for the same case as Fig. 3.21: (a) electric-field-driven velocity flux; (b) collisional flux; and (c) combined fluxes are shown. Note that (a) dominates (b) for nearly all positive $v_\parallel$.

$v_\parallel 0 \longrightarrow$

3.3.2. Scanning charge-exchange analyzer diagnostic for tokamaks

Tokamak fusion plasmas are generally considered to be fully ionized. However, a dilute population of neutral atoms usually coexists with the plasma in the confinement region. These neutrals arise from neutral beam sources used to heat the plasma and from molecular hydrogen dissociation in the edge plasma due to refluxing at the limiter or wall. Driven by these sources, the velocity distribution of these primary neutrals and their progeny evolves through ionization and further charge exchange, and through free streaming along straight-line orbits.

The average velocity of a neutral (with the exception of the primary beam neutrals) is typically comparable to that of the ambient ions within a charge-exchange mean free path of the neutral's position. The time scale on which the neutral distribution relaxes to a quasi-steady state is thus a straight-line thermal transit time across the plasma

$$\tau_{n_0} \sim \frac{a}{v_{th}} \sim \frac{3}{\sqrt{T_i \, (\text{keV})}} \, \mu s.$$

In a magnetic confinement device, this must be much faster than the fastest operant process working to change the ion distributions; i.e., slowing down, pitch-angle diffusion, and banana center diffusion. The neutral distribution can thus be viewed as determined by the ion distribution in the sense that following an initial transient the neutrals evolve quasistatically tracking the ions. These considerations suggest that measurement of the neutral spectrum at the plasma edge be used to monitor the ion distribution within the plasma. The presence of charge-exchange analyzer diagnostics on current experiments further motivated the construction of a numerical diagnostic for the bounce-averaged Fokker–Planck code. In this manner, it is possible not only to simulate the evolution of the plasma, but also the measurements taken, directly; the intent being that the diagnostic serve as a check to verify the adequacy of the model.

It is essential to the success of the diagnostic as a calibration tool that some realism be exercised in generating the background neutral distributions. However, the spectra of high-energy neutrals appearing at the charge-exchange analyzer port are quite insensitive to the details of the neutral distribution within the plasma: radially resolved profiles of background neutral density and average energy are generally sufficient for accurate representation. To generate these profiles, we have used a neutral transport code developed by Burrell (1978) and modernized by Stockdale (1983). The neutral transport calculation is performed in cylindrical geometry and includes the effects of wall (diffuse) reflection and volume neutral sources (neutral beam trapping, recombination). The details of the calculation have been treated elsewhere and will not be discussed further here. The results are shown in Figs. 3.23 and 3.24. Figure 3.23

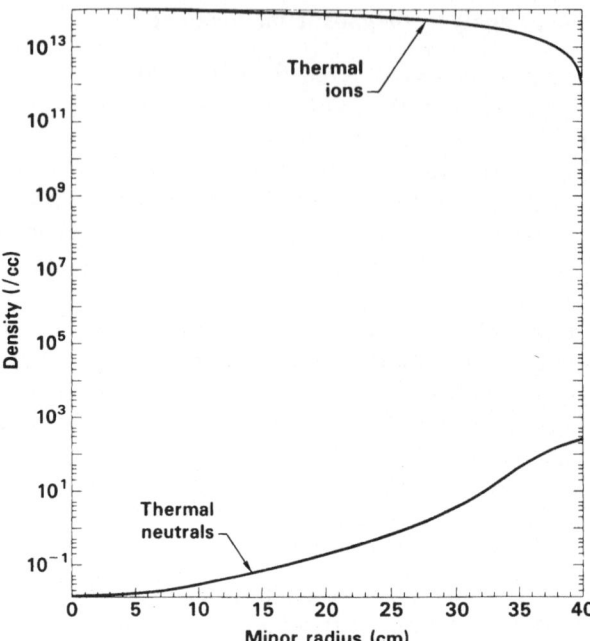

Figure 3.23. Profiles of background neutral density and plasma density. A neutral transport code developed by Burrell (1978) and modernized by Stockdale (1983) is used for the neutral calculation. The neutral transport calculation is performed in cylindrical geometry and includes the effects of wall (diffuse) reflection and volume neutral sources (neutral beam trapping, recombination).

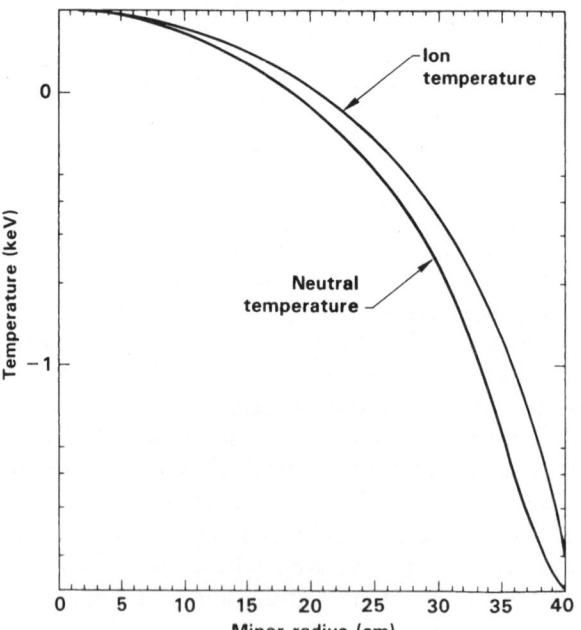

Figure 3.24. Profiles of average neutral energy and plasma temperature. The plasma profiles in this figure and Fig. 3.23 are preset, the neutral profiles are computed.

depicts the plasma and neutral density profiles as functions of minor radius; Fig. 3.24 depicts the profiles of ion and neutral average energy.

The charge-exchange event leaves the velocities of the products essentially unchanged from their values prior to the interaction. This is the reason detection of neutrals can provide information about the ion velocity distribution. Unfolding the charge-exchange data gathered by the experimental diagnostic to recover the ion distribution is effectively the inverse of the calculation done in the numerical diagnostic where the ion distribution is known prior to the neutral spectra. Both calculations involve integrating the interaction (charge-exchange and ionization) operator over the neutral orbits from their birthplace to the detector. In one aspect the two exercises differ. The experimental apparatus generates a radially integrated composite signal, a line-averaged quantity. The code provides radial resolution.

In the code calculation, we limit our model to the detection of neutrals born on, and traveling in, the equatorial plane of the torus. This simplifies the geometry and, moreover, in most cases also coincides with the experimental setup. The procedure uses the calculated small banana-width ion distribution as the ensemble of particles (in the finite banana-width case) all of which orbit the same bounce-averaged flux surface, $\Psi(r) = \langle\!\langle \Psi \rangle\!\rangle = \bar\Psi$. Each $(v_0, \theta_0, \bar\Psi)$-orbit crosses the equatorial plane twice in a full bounce (or transit) period.

In order that a charge-exchange event be detectable, it must coincide with an equatorial plane crossing. The position of the crossing is obtained through the solution of (3.1.142), the condition for conservation of canonical toroidal angular momentum. An iterative scheme is employed to accomplish this. The equation is first expressed as a transcendental relation in $r = \Psi^{-1}(\Psi)$ at equatorial plane crossing. Solution then yields the minor radius position of the guiding center at those bounce-phase points. Co-passing particles cross the equatorial plane outside the (toroidal) flux surface $\Psi = \bar\Psi$ at poloidal angle $\theta_p = 0$ and inside at $\theta_p = \pi$; for counter-passing particles the case is reversed; trapped particles cross outside on the co-moving leg and inside on the counter-moving leg, both crossings occurring for $\theta_p = 0$.

Figure 3.25 depicts the outcome of this calculation as contour maps in the midplane velocity system. All gyro-center orbits which pierce the midplane velocity space (viewed as a Poincaré map of the orbits) along a given contour will cross the equatorial midplane of the torus (a physical plane) at the same minor radius. In this sense, each contour can be viewed as the image in midplane velocity space of the locus of points in physical space all of a given minor radius at which detectable charge-exchange events can take place. Coincidentally, the contour corresponding to minor radius $r = a$ is also sometimes referred to as the "loss cone boundary". The two figures differ in the passing orbit region of velocity space: (a) shows the outboard equatorial crossings; and (b) shows the inboard equatorial crossings. This distinction is unnecessary for trapped orbits.

The detection of neutrals born at these equatorial plane crossing occurs

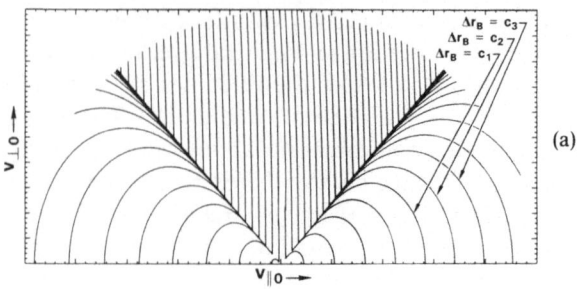

(a)

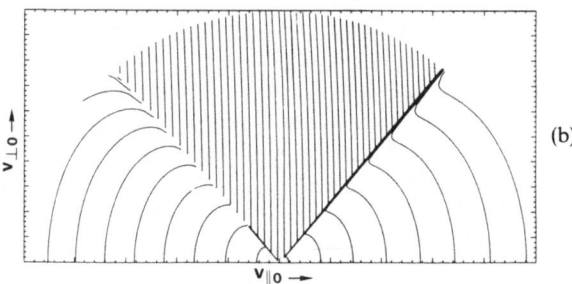

(b)

Figure 3.25. Equatorial plane crossings. All gyro-center orbits which pierce the midplane velocity space (viewed as a Poincaré map of the orbits) along a given contour will cross the equatorial plane of the torus (a physical plane) at the same minor radius. In this sense, each contour can be viewed as the image in midplane velocity space of the locus of points in physical space all of a given minor radius at which detectable charge-exchange events can take place. The contour corresponding to minor radius $r = a$ is also referred to as the "loss cone boundary". The two figures differ in the passing orbit region of velocity space: (a) shows the outboard equatorial crossings; and (b) shows the inboard equatorial crossings. This distinction is unnecessary for trapped orbits (cf. Section 3.3.2).

along the line of sight corresponding to the straight-line orbit of the neutral subsequent to its birth. For purposes of the orbit integration, the charge-exchange event is assumed to occur at the position of the ion gyro-center.

The line of sight is parametrized by R_T, the major radius at which it is tangent to a (toroidal) flux surface. This tangency radius, R_T, is related to the pitch-angle of the gyro-motion at equatorial plane crossing. $\theta_{ep}(v_0, \theta_0)$ and to the pitch-angle of the magnetic field lines, β. A short exercise in geometry demonstrates that

$$R_T = (R_0 + r + \Delta r_B)\cos\theta_{ep}/\cos\beta,$$

where R_0 is the major radius, r is the flux surface minor radius, $\Delta r_B(v_0, \theta_0)$ is the finite banana width radial excursion at equatorial plane crossing, and β is given by

Figure 3.26. Contours of
equatorial R_T on the midplane
velocity space. Those orbits
which underlie a given contour
correspond to ions which upon
charge exchange at equatorial
plane crossing generate neutrals
moving along lines of sight with
equal R_T. There are no contours
near $\theta_0 = 0, \pi$, since those orbits
have insufficient $v_\perp$ to produce
neutrals whose motion remains
in the equatorial plane.

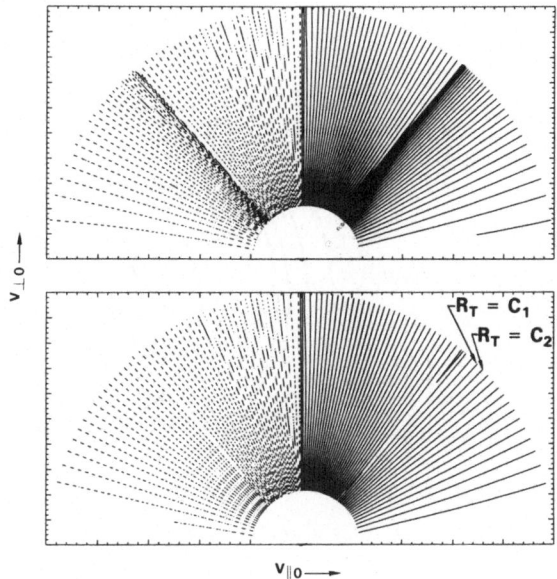

$$\cos \beta = \frac{B_T}{|B|} = \hat{\mathbf{e}}_\zeta \cdot \hat{b},$$

with $\theta_{ep} \geq \beta$. Orbits with insufficient $v_\perp$, according to this last relation, are not
detectable by the analyzer, and the resultant neutral velocity is not directed in
the equatorial plane. Figure 3.26 shows contours of equatorial R_T on the
midplane velocity space. Those orbits which underlie a given contour correspond to ions which upon charge exchange at equatorial plane crossing generate neutrals moving along lines of sight with equal R_T. Note there are no
contours near $\theta_0 = 0, \pi$, for the reasons mentioned previously. Figure 3.27
shows the geometrical arrangement of the scanning charge-exchange analyzer
(in physical space).

The neutral flux is attenuated along the line of sight to the detector by
interaction with the ambient plasma according to

$$\Phi = \Phi_0 \exp\left(-\int \frac{dl}{\lambda(l)}\right).$$

The interaction mean free path $\lambda(l)$ viewed as a function of orbit position is
given by

$$v\lambda^{-1} = n_i(\langle \sigma v \rangle_{cx} + \langle \sigma v \rangle_{ii}) + n_e \langle \sigma v \rangle_{ei}.$$

Here the charge-exchange, ion, and electron-impact ionization reaction rates
are averaged over a Maxwellian ambient background characterized by (radially resolved) profiles preset as mentioned earlier. λ is a function of neutral

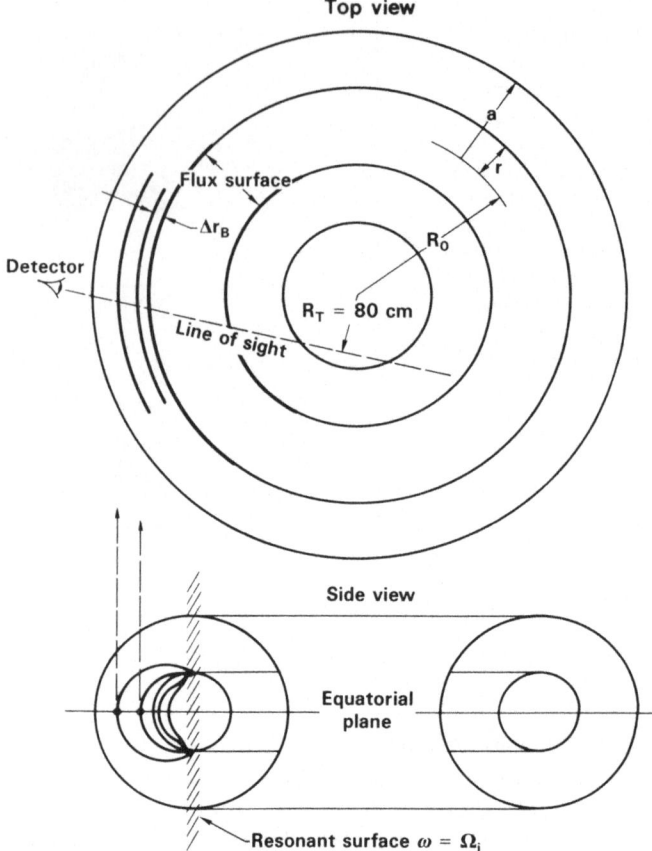

Figure 3.27. Scanning charge exchange analyzer geometry viewed in the equatorial plane of the tokamak.

energy and orbit position. The attenuation integral calculation is done as a setup for each meshpoint, stored in a "sensitivity" array, and used repeatedly to interpret current values of the ion distributions at meshpoints as neutral spectra in R_T. The calculation is arranged so different flux surfaces can be considered through parallel calculations in consistently reproduced ambient conditions. This allows direct correlation with the experimental diagnostic through summation. In view of Figs. 3.26 and 3.25 it is clear neutrals born of a given (average) flux surface ensemble, whose subsequent trajectories are associated with a given R_T, each suffer different attenuation on their journey to the analyzer, though all enter the analyzer from the same direction.

Figure 3.28 shows the spectral neutral flux at the charge-exchange analyzer port as a function of R_T for various energies. These spectra are computed by

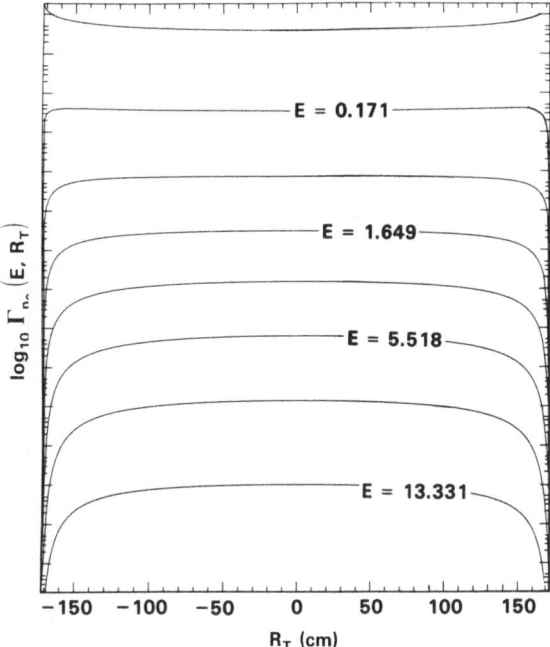

Figure 3.28. Spectral neutral flux at the charge-exchange analyzer port as a function of R_T for various energies. These spectra are computed by the neutral transport package (Stockdale, 1983). That calculation presumes a cylindrically symmetric plasma; there are neither trapped particle effects nor finite banana width effects operant in the model. The spectra are the result of charge-exchange events which occur in the equatorial plane between the ambient neutrals and the local Maxwellian ion background, attenuated along the line of sight to the charge-exchange analyzer port, and (the result) integrated over the minor radius. At large R_T the neutrals entering the analyzer are neutrals which have been born at the plasma periphery, and therefore their spectra are characteristic of a lower average energy.

the neutral transport package written by Burrell (1978) mentioned previously. That calculation presumes a cylindrically symmetric plasma; there are neither trapped particle effects nor finite banana-width effects operant in the model The spectra are the result of charge-exchange events which occur in the equatorial plane between the ambient neutrals and the local Maxwellian ion background, attenuated along the line of sight to the charge-exchange analyzer port, and (the result) integrated over the minor radius. At large R_T the neutrals entering the analyzer are neutrals which have been born at the plasma periphery, and therefore their spectra are characteristic of a lower average energy.

Figure 3.29 shows the results of a related calculation. Here the thermal neutral spectra arise through charge exchange of ions all in the same (average) flux surface ensemble with ambient neutrals: finite banana-width effects are

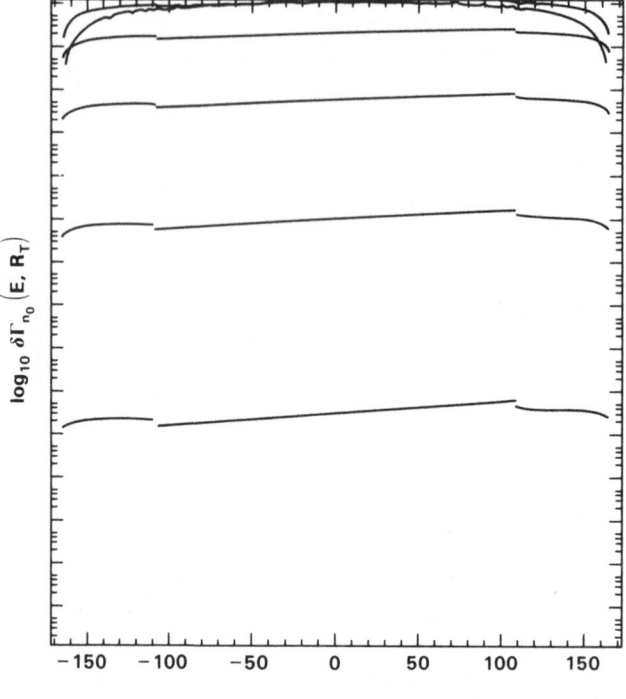

Figure 3.29. Thermal neutral spectra arising through charge exchange of ions all in the same (average) flux surface ensemble with ambient neutrals: finite banana-width effects are included. Charge-exchange events which occur on the (outer) co-moving leg of trapped orbits produce larger neutral fluxes at the analyzer than those which occur on the (inner) counter-moving leg. This is a consequence of both larger neutral density at larger minor radii and the relatively smaller attenuation integral associated with outboard events.

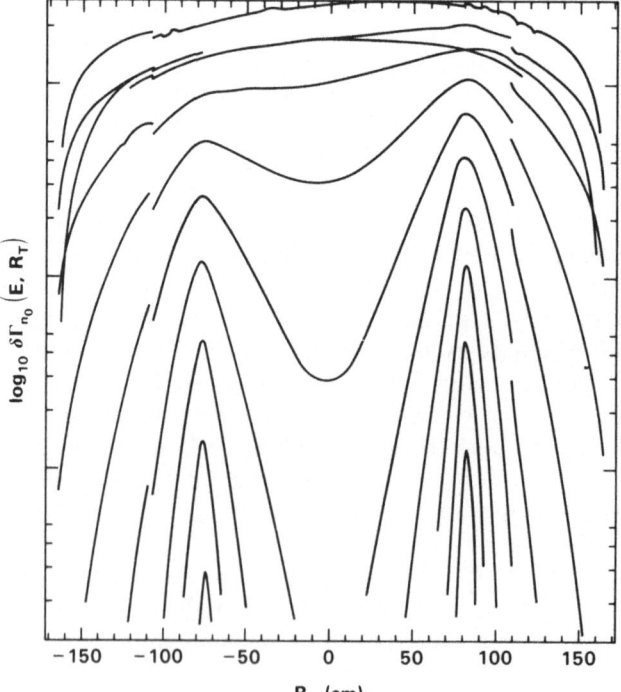

Figure 3.30. The same case as Fig. 3.29 after the application of fundamental minority species heating RF excitation. The distribution mapped through this diagnostic procedure is shown in Fig. 3.8. The finite banana-width equivalent to Fig. 3.28 would be a sum of results like Fig. 3.30, one for each flux surface.

included. For example, notice that charge-exchange events which occur on the (outer) co-moving leg of trapped orbits produce larger neutral fluxes at the analyzer than those which occur on the (inner) counter-moving leg. This is a consequence of both larger neutral density at larger minor radii and the relatively smaller attenuation integral associated with outboard events. Figure 3.30 represents the same case as Fig. 3.28 after the application of fundamental minority species heating RF excitation. The distribution mapped through this diagnostic procedure is shown in Fig. 3.8. The finite banana-width equivalent to Fig. 3.28 would be a sum of results like Fig. 3.30, one for each flux surface. Kaita *et al.* (1983) have demonstrated such RF signatures in data taken on PLT.

Appendix 3A. Coefficients of the Bounce-Averaged Operator

As a consequence of the requirement that the Fokker–Planck operator conserve particles (collisions simply alter the local phase space density) we can write

$$\frac{\partial N}{\partial t} = 0, \tag{3A.1}$$

where N, the number of particles whose gyro-centers are associated with a given magnetic field line on the interval $-s_{mx} < s < s_{mx}$, is given by the integral

$$N = \int_0^{s_B} ds \int d^3v_0 \frac{dn}{ds}, \tag{3A.2}$$

where

$$\frac{dn}{ds}(v, \theta, s) = \frac{v_{\|0}}{v_\|} \mathscr{F}_0 \tag{3A.3}$$

is the incremental phase space density of gyro-centers between s and $s + ds$ which pierce the midplane velocity space at v_0, θ_0. The velocity integral is taken over the entire (infinite) domain, and we have used the symbols s_B and s_{mx} to denote the arclength along a field line from the field minimum in a symmetric magnetic well to a gyro-center bounce point, and to the field maximum, respectively. The total number of gyro-centers in the ensemble, N, is evenly distributed in bounce-phase ϕ_B, but not in s; hence the ratio of parallel velocities in (3A.3).

Inserting (3A.2) and (3A.3) in (3A.1) we have

$$0 = \frac{\partial N}{\partial t} = \frac{\partial}{\partial t} \int_0^{s_B} ds \int d^3v_0 \frac{v_{\|0}}{v_\|} \mathscr{F}_0 = \int d^3v_0 \frac{\partial}{\partial t} \lambda \mathscr{F}_0, \tag{3A.4}$$

where $\lambda = v_{\|0} \tau_B$. Equation (3A.4) can be recast by using the local Fokker–

Planck operator to replace $\partial \mathscr{F}/\partial t$ as

$$
\begin{aligned}
0 = \frac{\partial N}{\partial t} &= \frac{\partial}{\partial t} \int_0^{s_{mx}} ds \int d^3v \frac{d^3v_0}{d^3v} \frac{v_{\|0}}{v_\|} \mathscr{F} \\
&= \int_0^{s_{mx}} \frac{ds}{\psi} \int d^3v \frac{\partial \mathscr{F}}{\partial t} = \int_0^{s_{mx}} \frac{ds}{\psi} \int d^3v\, \nabla_v \cdot \hat{\Gamma} \\
&= \int dv_0^3 \int_0^{s_B} \frac{ds}{\psi} \frac{d^3v}{dv_0^3} \nabla_v \cdot \hat{\Gamma} = \int d^3v_0\, \lambda \langle\!\langle \nabla_v \cdot \hat{\Gamma} \rangle\!\rangle.
\end{aligned}
\tag{3A.5}
$$

From (3A.5) we may infer that the integrand in the last integral may be represented as the divergence of a flux in the coordinates v_0, θ_0, i.e.,

$$
0 = \int d^3v_0\, \lambda \langle\!\langle \nabla_v \cdot \hat{\Gamma} \rangle\!\rangle = \int d^3v_0\, \nabla_{v_0} \cdot \hat{\Gamma}_0.
\tag{3A.6}
$$

If this were not the case, it would not be possible to write the bounce-averaged Fokker–Planck equation in conservative form: Orbit averaging must preserve this invariance property of the system. The entity conserved before gyro-phase averaging was the number of particles in the ensemble $\hat{f}$. After gyro-phase averaging, the entity conserved became the number of gyro-centers in the ensemble f. After bounce-averaging the invariant quantity is the number of bounce centers (including transit centers) in the ensemble $\mathscr{F}_0$.

The proof of (3A.5) depends on the fact that $\hat{\mathbf{e}}_\theta \cdot \Gamma(s_B) = 0$, stemming from the requirement that $\mathscr{F}$ be symmetric about $\pi/2$ in a neighborhood of $\pi/2$ (and for $s \neq s_{mx}$).

The quantities $\mathscr{G}$, $\mathscr{H}$ defined in (3.1.57) and (3.1.58) can also be written

$$
\mathscr{G} = v^2 \hat{\mathbf{e}} \cdot \hat{\Gamma}; \qquad \mathscr{H} = v \sin\theta \hat{\mathbf{e}}_\theta \cdot \hat{\Gamma}.
\tag{3A.7}
$$

The bounce-averaged divergence appearing in (3A.5) thus expands to

$$
\begin{aligned}
\langle\!\langle \nabla_v \cdot \hat{\Gamma} \rangle\!\rangle &= \frac{1}{\tau_B} \int_0^{s_B} \frac{ds}{v_\|} \left[\frac{1}{v^2} \frac{\partial}{\partial v} (v^2 \hat{\mathbf{e}}_v \cdot \hat{\Gamma}) + \frac{1}{v^2 \sin\theta} \frac{\partial}{\partial \theta} (v \sin\theta \hat{\mathbf{e}}_\theta \cdot \hat{\Gamma}) \right] \\
&= \frac{1}{\lambda v_0^2 \sin\theta_0} \int_0^{s_B} ds \left[\frac{\partial}{\partial v_0} \left(\frac{\cos\theta_0}{\sqrt{\psi}\,\cos\theta} v^2 \sin\theta \hat{\mathbf{e}}_v \cdot \hat{\Gamma} \right) \right. \\
&\qquad \left. + \frac{\cos\theta_0}{\sqrt{\psi}\,\cos\theta} \frac{\partial\theta_0}{\partial\theta} \frac{\partial}{\partial\theta_0} (v \sin\theta \hat{\mathbf{e}}_\theta \cdot \hat{\Gamma}) \right] \\
&= \frac{1}{\lambda v_0^2 \sin\theta_0} \left[\frac{\partial}{\partial v_0} \int_0^{s_B} ds \frac{\cos\theta_0}{\sqrt{\psi}\,\cos\theta} v^2 \sin\theta \hat{\mathbf{e}}_v \cdot \hat{\Gamma} \right. \\
&\qquad \left. + \frac{\partial}{\partial\theta_0} \int_0^{s_B} \frac{ds}{\psi} (v \sin\theta \hat{\mathbf{e}}_\theta \cdot \hat{\Gamma}) \right] \\
&= \frac{1}{\lambda} \nabla_{v_0} \cdot \hat{\Gamma}_0,
\end{aligned}
\tag{3A.8}
$$

where we have used (3A.6) and

$$\frac{\partial}{\partial \theta} = \frac{\partial \theta_0}{\partial \theta} \frac{\partial}{\partial \theta_0} + \frac{\partial v_0}{\partial \theta} \frac{\partial}{\partial v_0} = \frac{\cos \theta}{\sqrt{\psi} \cos \theta_0} \frac{\partial}{\partial \theta_0}.$$

As in (3.1.55), we define the bounce-averaged flux components $\mathcal{G}_0$, $\mathcal{H}_0$ so that

$$\nabla_{v_0} \cdot \hat{\Gamma} = \frac{1}{v_0^2} \frac{\partial}{\partial v_0} \mathcal{G}_0 + \frac{1}{v_0^2 \sin \theta_0} \frac{\partial}{\partial \theta_0} \mathcal{H}_0, \tag{3A.9}$$

where

$$\mathcal{G}_0 = \int ds \frac{\cos \theta_0}{\cos \theta} v_0^2 \hat{e}_v \cdot \hat{\Gamma} = \lambda \langle\!\langle \mathcal{G} \rangle\!\rangle, \tag{3A.10}$$

and where

$$\mathcal{H}_0 = \int \frac{ds}{\psi} \sin \theta \hat{e}_\theta \cdot \hat{\Gamma} = \lambda \left\langle\!\left\langle \frac{\cos \theta}{\psi \cos \theta_0} \mathcal{H} \right\rangle\!\right\rangle. \tag{3A.11}$$

Identification of the bounce-averaged coefficients defined in (3.1.55) using (3A.10) and (3A.11) yields the relations (3.1.71), and these are valid for any conservative operator of the form (3.1.50).

Appendix 3B. Boundary Layer Diagnostic

In order to implement the numerical solution Fokker–Planck equations in nonuniform magnetic fields in conservation form (cf. (3.1.54)–(3.1.55)) it is necessary to compute integrals of the form

$$N = \int d^3 v_0 \, v_{\|0} \tau_B \mathcal{F}_0 = 2\pi \int dv_0 \, v_0^3 \int d\mu_0 \, \mu_0 \tau_B \mathcal{F}_0 \equiv 2\pi \int dv_0 \, v_0^3 I. \tag{3B.1}$$

Here we have used the notation $\mu_0 = \cos \theta_0$ and τ_B represents the bounce-period, or transit-period, for gyro-centers in a (symmetric) magnetic well. The computation is complicated by the fact that τ_B diverges logarithmically in the neighborhood of the boundary separating trapped and circulating particle orbits. In magnetic wells of the form we consider here, symmetric wells with no ambipolar potential, this boundary is represented by the lines $\mu_0 = \pm \mu_{0_T}$: in the more general case, the boundary and the velocity space orbit topology (cf. Section 3.1.3) may be considerably more complicated, yet there still arise integrals of the form (3B.1) with the same inherent difficulty.

The integral I in (3B.1) is well defined and finite and an algorithm for evaluating it accurately and efficiently is essential for some problems. Of particular concern are problems in which trapping and detrapping rates are an issue. In such cases we are concerned with treating fluxes between circulating and trapped orbits so as to assuure that physical effects dominate numerical inaccuracies in a neighborhood of the boundary.

Let us limit the discussion which follows to a thin layer straddling the boundary between trapped and circulating orbits. Further, let us define the extent of this layer by $\mu_0^- < \mu_0 < \mu_0^+$, where $\mu_0^- < \mu_{0_r} < \mu_0^+$. By (3.1.119), and the discussion following, we can approximate $\mathscr{F}_0$ as piecewise linear in μ_0 on this domain and thus take

$$\mathscr{F}_0 = \begin{cases} \alpha^- + \mu_0 \beta^-, & \mu_0^- < \mu_0 < \mu_{0_r}, \\ \alpha^+ + \mu_0 \beta^+, & \mu_{0_r} < \mu_0 < \mu_0^+. \end{cases} \tag{3B.2}$$

The coefficients in (3B.2) are independent of μ_0. Also define

$$I_{bl} = \int_{\mu_0^-}^{\mu_0^+} d\mu_0 \, \mu_0 \tau_B \mathscr{F}_0 \tag{3B.3}$$

and for convenience let $\mathscr{F}_0(\mu_{0_r}) \equiv \mathscr{F}_{0_r}$, $\mathscr{F}_0(\mu_0^\pm) \equiv \mathscr{F}_0^\pm$. Further, let

$$\mu_0 = \mu_{0_r}(1 + \delta) \tag{3B.4}$$

define the quantity δ which ranges over the values $-\Delta < \delta < \Delta$ within the thin layer $\mu_0^- < \mu_0 < \mu_0^+$. It is our intention, after a few preliminary preparations, to integrate over this δ within the defined range to get an approximation for the integral (3B.3).

To get an analytic representation for I_{bl} it is necessary to choose $\psi(s)$, a magnetic field, to obtain an expression for τ_B in the neighborhood of $\mu_0 = \mu_{0_r}$. To illustrate the calculation, let $\psi(s)$ be given by

$$\psi(s) = \frac{1 + \varepsilon}{1 + \varepsilon \cos \pi s/s_{mx}}, \tag{3B.5}$$

where ε is the inverse aspect ratio of an axisymmetric torus, $\varepsilon = r/R_0$, and $\pi s/s_{mx}$ is the poloidal angle. The bounce time is then given by

$$\tau_B = \int_0^{s_B} \frac{ds}{v_\parallel} = \int_0^{s_B} \frac{ds}{v_0\sqrt{1 - \psi(s)(1 - \mu_0^2)}}. \tag{3B.6}$$

As previously, we have used the symbols s_B and s_{mx} to denote the arclength along a field line from the field minimum in a symmetric magnetic well to a gyro-center bounce point, and to the field maximum, respectively. Using the identity $\cos \phi = 1 - 2 \sin^2(\phi/2)$ and the transformation $\zeta = \pi s/2s_{mx}$, (3B.6) can be recast as

$$\tau_B = \frac{2s_{mx}}{\pi v_0 \mu_0} \int_0^{\zeta_B} d\zeta \, \frac{\sqrt{1 - \mu_{0_r}^2 \sin^2 \zeta}}{\sqrt{1 - (\mu_{0_r}^2/\mu_0^2) \sin^2 \zeta}}, \tag{3B.7}$$

where

$$\zeta_B = \begin{cases} \frac{1}{2}\arccos(1 - 2\mu_0^2/\mu_{0_r}^2), & \mu_0^2 \le \mu_{0_r}^2, \\ \frac{1}{2}\pi, & \mu_0^2 > \mu_{0_r}^2. \end{cases} \tag{3B.8}$$

By expanding the numerator in the integrand of (3B.7), a series expansion for τ_B can be generated, each of whose terms is an integral which can be represented in terms of elliptic integrals

$$\tau_B = \frac{2s_{mx}}{\pi v_0 \mu_0} \sum_{m=0}^{\infty} \alpha_m \mu_{0T}^{2m} \int_0^{\zeta_B} d\zeta \frac{\sin^{2m} \zeta}{\sqrt{1 - (\mu_{0_T}^2/\mu_0^2) \sin^2 \zeta}}, \tag{3B.9}$$

where the coefficients α_m are

$$\alpha_0 = 1, \quad \alpha_1 = -\tfrac{1}{2}, \ldots, \alpha_m = \frac{2m - 3}{2m} \alpha_{m-1}. \tag{3B.10}$$

Using the fact that $\sin^2 \zeta_B = \mu_0^2/\mu_{0_T}^2$, the series (3B.9) can be evaluated recursively as

$$\tau_B = \frac{2s_{mx}}{\pi v_0 \mu_0} \sum_{m=0}^{\infty} \alpha_m \mu_{0T}^{2m} J_{2m}, \tag{3B.11}$$

where

$$J_{2m} = \frac{1}{2m - 1}\left((2m - 2)\left(1 + \frac{\mu_0^2}{\mu_{0_T}^2}\right) J_{2m-2} - (2m - 3)\frac{\mu_0^2}{\mu_{0_T}^2} J_{2m-4}\right). \tag{3B.12}$$

To begin the recursion, values for J_0 and J_2 are requried: These functions are

$$J_0 = \begin{cases} K\left(\dfrac{\mu_{0_T}^2}{\mu_0^2}\right), & \mu_0^2 \geq \mu_{0_T}^2, \\[2ex] \dfrac{\mu_0}{\mu_{0_T}} K\left(\dfrac{\mu_0^2}{\mu_{0_T}^2}\right), & \mu_0^2 \leq \mu_{0_T}^2, \end{cases} \tag{3B.13}$$

and

$$J_2 = \begin{cases} \dfrac{\mu_0^2}{\mu_{0_T}^2}\left[K\left(\dfrac{\mu_{0_T}^2}{\mu_0^2}\right) - E\left(\dfrac{\mu_{0_T}^2}{\mu_0^2}\right)\right], & \mu_0^2 \geq \mu_{0_T}^2, \\[2ex] \dfrac{\mu_0}{\mu_{0_T}}\left[K\left(\dfrac{\mu_0^2}{\mu_{0_T}^2}\right) - E\left(\dfrac{\mu_0^2}{\mu_{0_T}^2}\right)\right], & \mu_0^2 \leq \mu_{0_T}^2, \end{cases} \tag{3B.14}$$

where K, E are complete elliptic integrals.

Using a polynomial approximation for K, E near modulus 1 and expanding the modulus in δ, the contribution to I from the boundary layer $-\Delta \leq \delta \leq \Delta$ can be written as

$$I_{bl} = \frac{2s_{mx}}{\pi v_0} \sum_{m=0}^{\infty} \alpha_m \mu_{0_T}^{2m+1} \left\{ \int_{-\Delta}^0 d\delta\, (\alpha^- + \beta^- \mu_{0_T}(1 + \delta)) J_{2m}^-(\delta) \right. \\ \left. + \int_0^{\Delta} d\delta\, (\alpha^+ + \beta^+ \mu_{0_T}(1 + \delta)) J_{2m}^+(\delta) \right\}, \tag{3B.15}$$

where $J_{2m}^{\pm}$ involve the polynomial approximations to the elliptic integrals through (3B.12), (3B.13), and (3B.14). Performing the integrals and working to first order in Δ, we find

$$I_{bl} = \frac{4 s_{mx} \mathscr{F}_{0_T} \Delta}{\pi v_0} \left[\mu_{0_T} \sqrt{1 - \mu_{0_T}^2} (k' - k''(\ln 2\Delta - 1)) \right.$$
$$\left. - \sum_{m=0}^{\infty} \alpha_m \mu_{0_T}^{2m+1} \beta_m + O(\Delta^2 \ln \Delta) \right], \tag{3B.16}$$

where k' and k'' are coefficients in the expansion for $K(x) = k' - k'' \ln x + O(x)$, and β_m is given by the recursion relation

$$\beta_m = \frac{2(2m - 2)\beta_{m-1} - (2m - 3)\beta_{m-2}}{2m - 1} \tag{3B.17}$$

with $\beta_0 = 0$ and $\beta_1 = 1$.

To provide a more general field model, we choose a cubic spline fit to $\psi(s)$ on successive internal intervals of s, $0 < s_1 < s_2 < \cdots < s_{L-1} = s^* < s_{mx}$. Choose ψ to be quadratic on the end intervals such that $\psi'(0) = \psi'(s_{mx}) = 0$, $\psi(0) = 1$, and $\psi(s_{mx}) = \psi_{mx}$. The procedure is entirely analogous to that followed in the previous case with the exception of one detail: here we will represent the bounce time $\tau_B(\mu_0)$ in the neighborhood of $\mu_0 = \mu_{0_T}$ as a sum of two contributions, one being convergent and the other being logarithmically divergent, but representable in closed form to requisite order. This allows the integral I_{bl} to be performed as in the previous treatment.

Therefore, let

$$\tau_B = \tau_1 + \tau_2, \tag{3B.18}$$

where

$$\tau_1 = \frac{1}{v_0} \int_0^{s^*} \frac{ds}{\mu(\mu_0)} \tag{3B.19}$$

and

$$\tau_2 = \frac{1}{v_0} \int_{s^*}^{s_B(\mu_0)} \frac{ds}{\mu(\mu_0)}. \tag{3B.20}$$

Since the integrand of I is to be represented near $\mu_0 = \mu_{0_T}$ we expand τ_1 as a Taylor series near this point

$$\tau_1(\mu_0) = \tau_1(\mu_{0_T}) + \tau_1'(\mu_{0_T})(\mu_0 - \mu_{0_T}).\ldots \tag{3B.21}$$

Each of the terms of this series is calculable numerically—there are no divergent terms. Furthermore, by assuring that the boundary layer is chosen thin enough so that within the layer

$$\mu_0^2 = (\mu_{0_T} \pm \Delta)^2 > (\mu_0^{*2}) = 1 - \frac{1}{\psi^*}, \tag{3B.22}$$

where we have abbreviated $\psi^* = \psi(s^*)$. The quantity μ_0^{*2} will be recognized as the cosine of the midplane pitch-angle characterizing the orbit which bounces at s^*. It can be shown that the second term in (3B.21) contributes to I only at $O(\Delta^2)$.

The second contribution to τ_B, τ_2, becomes

$$\tau_2 = \frac{h\alpha}{2v_0\mu^*} \ln \frac{|\alpha + 1|}{|\alpha - 1|}, \tag{3B.23}$$

where

$$\begin{aligned}
h &= s_{mx} - s^*, \\
\mu^{*2} &= 1 - \psi^*(1 - \mu_0^2), \\
\alpha^2 &= \left(1 - \frac{\mu_{mx}^2}{\mu^{*2}}\right)^{-1}, \quad \text{and} \\
\mu_{mx}^2 &= 1 - \psi_{mx}(1 - \mu_0^2).
\end{aligned} \tag{3B.24}$$

It remains simply to insert expressions (3B.21) and an approximation for (3B.23) in the boundary layer into (3B.3) using (3B.2) for $\mathscr{F}_0$, then perform the integrals. The result including terms to $O(\Delta)$, $O(\Delta \ln \Delta)$ can be integrated numerically.

Appendix 3C. Tangent Resonance Phenomena

We wish to evaluate bounce-averages such as $\langle\!\langle S_n \sqrt{\psi}\, e^{i\Psi} \rangle\!\rangle$ (cf. 3.1.103)–(3.1.106)) in the neighborhood of tangent resonance. We represent these trajectory integrals as

$$I = \langle\!\langle \Pi(\tau)\, e^{i\Psi(\tau)} \rangle\!\rangle, \tag{3C.1}$$

where

$$\Psi(\tau) = -\int_0^\tau d\tau' \, (\omega - k_\parallel v_\parallel - m\Omega)(\tau') = -\int_0^\tau d\tau' \, v(\tau'), \tag{3C.2}$$

and Π is a slowly varying function of τ in the sense that $\dot{\Pi}/\Pi \sim \omega_B \ll \Omega$. Near the stationary phase points we expand the (rapidly oscillating phase) factor Ψ as

$$\Psi = \Psi_R + v\Delta\tau + \dot{v}\frac{\Delta\tau^2}{2!} + \ddot{v}\frac{\Delta\tau^3}{3!} + \cdots, \tag{3C.3}$$

$\Delta\tau$ being measured from the stationary phase point, $v(\tau_i) = 0$ (i.e., $\Delta\tau = \tau - \tau_i$). Placing (3C.3) in (3C.1) we can write

$$I = \sum_i \frac{1}{\tau_B} \int_{-D\tau}^{+D\tau} d\Delta\tau \, \Pi(\tau_i + \Delta\tau) \exp\left[i\left(\Psi_R + v\Delta\tau + \dot{v}\frac{\Delta\tau^2}{2!} + \ddot{v}\frac{\Delta\tau^3}{3!}\right)\right]. \tag{3C.4}$$

Now we pass to the limit $D\tau \to \infty$:

$$\lim_{D\tau \to \infty} I = \sum_i \frac{1}{\tau_B} \Pi(\tau_i) \, e^{i\Psi_R(\tau_i)} \int_{-\infty}^{+\infty} d\Delta\tau \, \exp\left[i\dot{v} \frac{\Delta\tau^2}{2!} \right]$$

$$= \sum_i \frac{1}{\tau_B} \Pi(\tau_i) \, e^{i\Psi_R} \, e^{i\pi/4} \sqrt{\frac{2\pi}{\dot{v}}} \,. \tag{3C.5}$$

A simple modification to the result (3C.5) can be used to model the effects of gyro-phase decorrelation. Let the n-bounce interaction history in the case of a single isolated resonance be represented in the form

$$I_n = \Pi(\tau_R)|\tau_c|\Sigma_n, \qquad v(\tau_R) = 0 \tag{3C.6}$$

where Σ_n is defined recursively by

$$\Sigma_n = \Sigma_{n-1}\{\Phi(\delta\phi_n, 0) + \Phi(\delta\phi_n, 4\tau_B)e^{-i4\Psi_x}\} = S_n\Sigma_0, \tag{3C.7}$$

with

$$\Sigma_0 = e^{i(\Psi_R + \pi/4)}[\Phi(\delta\phi_0, 0) + \Phi(\delta\phi_0, 2\tau_R)e^{-2i(\Psi_R + \pi/4)}] + c.c. \tag{3C.8}$$

Here we have introduced the phase integrals

$$\Psi_R = \int^{\tau_R} d\tau \, v(\tau), \qquad \Psi_x = \int^{\tau_B} d\tau \, v(\tau),$$

and the random gyro-phase increments $\delta\phi_m$. In order to include the effects of collisions within this formalism, we introduce a gyro-phase diffusion kernel $\Phi(\delta\phi, \Delta\tau) = \exp(i\delta\phi)P(\delta\phi, \Delta\tau = \tau_1 - \tau_2)$ to describe the evolution during the time interval $\Delta\tau$ along a gyro-center trajectory of the probability distribution of cumulative random gyro-phase increments $\delta\phi$ due to pitch-angle scattering. Our aim, once having chosen a model for calculating Φ, will be to perform the ensemble average over increments $\delta\phi$ using the appropriate $\Delta\tau$.

To this end we require that P, the probability distribution of $\delta\phi$, have the following properties: (1) $\int d\delta\phi \, P = 1$, so that the kernel Φ is normalized; (2) $\int d\delta\phi \, \delta\phi \, P = 0$, the kernel is symmetric in $\delta\phi$; and (3) $\int d\delta\phi \, \delta\phi^2 \, P = \langle\Delta\phi^2\rangle(\Delta\tau) \sim \gamma(v_0, \theta_0)v_{ii}\Omega^2\Delta\tau^3$, the gyro-phase dispersion solves Langevin equations for the cumulative diffusion of the gyro-phase angle due to pitch-angle scattering (see for example Cohen et al. (1983)). The function γ is of order unity and depends on particulars of the magnetic field geometry. For purposes of illustration, we choose the distribution of the $\delta\phi_m$ to be Gaussian:

$$P(\delta\phi_m) = \frac{1}{\sqrt{\pi} \, D_m} e^{-(\delta\phi_m/D_m)^2},$$

where $D_m^2 = 2\langle\Delta\phi^2\rangle(\Delta\tau_m)$ will depend on the collisionality and the orbit parameters.

The coefficients represented in (3.1.103)–(3.1.106) involve products of the single bounce interaction with the entire prior history of the interaction, viz.

$$B_{0_{ql}} \sim \lim_{n \to \infty} I_0^* I_n = \Pi^2(\tau_R)\tau_c^2(\Sigma_0^* \Sigma_\infty + cc) \tag{3C.9}$$

Within the context of certain simplifying assumptions, we can perform the ensemble average over the distribution of random gyro-phase increments $\delta\phi$: Defining

$$\langle Q \rangle_m = \int_{-\infty}^{+\infty} d\delta\phi_m \Phi(\delta\phi_m, \Delta\tau)Q,$$

it can be shown that

$$\langle \Sigma_0^* \Sigma_n \rangle_0 = 2S_n[1 + e^{-D_0^2} - 2e^{-D_0^2} \sin 2\Psi_R].$$

For simplicity, let $D_m^2 \equiv m^3 D^2$, $\forall m > 0$ and $\Delta\tau_m = 4\tau_B$ so that the distribution of gyro-phase increments operant on any given bounce is that given by Cohen *et al.* (1983). Iterating the procedure outlined and using the recursion (3C.7), the gyro-phase decorrelated version of (3C.9) becomes

$$\langle B_{0_{ql}} \rangle_\infty = 2\Pi^2(\tau_R)\tau_c^2 \Upsilon[1 + e^{-D_0^2} - 2e^{-D_0^2} \sin 2\Psi_R], \tag{3C.10}$$

where coherence carried through prior bounces is represented by

$$\Upsilon = \sum_{k=0}^{\infty} \cos 4k\Psi_x e^{-k^3 D^2}.$$

In practice, D^2 is sufficiently large to wash out any prior bounce coherence in modern tokamaks (though marginally so in mirror end cells). In particular, Υ can never become singular for $D^2 > 0$. However, D_0^2 can quite generally be small for *single* bounce resonances which are not well separated, yet not close enough to ignore the gyro-phase diffusion.

For trajectories at tangent resonance $\dot{v}(\tau_i) = v(\tau_i) = 0$, $D_0^2 \to 0$, the quantity I as represented in (3C.5) diverges, gyro-phase diffusion is negligible, and the procedure becomes a bit more elaborate. In the constants-of-motion space proximity of tangent resonant orbits—recall that $v(\tau)$ is a gyro-center orbit function—we expand the eikonal Ψ about the gyro-center orbit point ϑ defined by $\dot{v}(\vartheta) = 0$. Now $v(\vartheta) \neq 0$ except exactly at tangent resonance, where $\vartheta = \tau_i$, and we have in place of (3C.4) that

$$I = \frac{1}{\tau_B} \int_{-D\tau}^{+D\tau} d\Delta\tau \, \Pi(\vartheta + \Delta\tau) \exp\left[i\left(\Psi_R + v\Delta\tau + \ddot{v}\frac{\Delta\tau^3}{3!}\right)\right]. \tag{3C.11}$$

This procedure is necessitated by the fact that near tangent resonance there are two gyro-resonances close enough to one another to be indistinct from one another. The method used in (3C.4)–(3C.5) thus must be generalized. In (3C.11), v and $\ddot{v}$ are evaluated at ϑ, and since Π is a slowly varying function of τ, we represent it in the interval $\vartheta - D\tau < \tau < \vartheta + D\tau$ as

$$\Pi(\tau) = \Pi(\vartheta) + \dot{\Pi}(\vartheta)\Delta\tau. \tag{3C.12}$$

Inserting (3C.12), (3C.11) may be written as

$$I = \frac{1}{\tau_B} \left\{ \Pi(\vartheta) \, e^{i\Psi_R(\tau)} \int_{-\infty}^{+\infty} d\Delta\tau \, \exp\left[i\left(\nu\Delta\tau + \ddot{\nu}\frac{\Delta\tau^3}{3!} \right) \right] \right.$$

$$\left. + \dot{\Pi}(\vartheta) \, e^{i\Psi_R(\vartheta)} \int_{-\infty}^{+\infty} d\Delta\tau \, \Delta\tau \, \exp\left[i\left(\nu\Delta\tau + \ddot{\nu}\frac{\Delta\tau^3}{3!} \right) \right] \right\}. \tag{3C.13}$$

In the following, we make use of the relation defining the Airy function (Abramowitz and Stegun, 1968):

$$\frac{2\pi}{\sqrt[3]{3a}} Ai\left(\pm\frac{x}{\sqrt[3]{3a}} \right) = \mathrm{Re} \int_{-\infty}^{+\infty} d\zeta \, e^{i(a\zeta^3 \pm x\zeta)}. \tag{3C.14}$$

From (3C.14) there follows directly that

$$\frac{d}{dx} \frac{2\pi}{\sqrt[3]{3a}} Ai\left(\pm\frac{x}{\sqrt[3]{3a}} \right) = \pm\frac{2\pi}{\sqrt[3]{3a^2}} Ai'\left(\pm\frac{x}{\sqrt[3]{3a}} \right)$$

$$= \mathrm{Re} \pm \int_{-\infty}^{+\infty} d\zeta \, i\zeta e^{i(a\zeta^3 \pm x\zeta)} \tag{3C.15}$$

from which there derives

$$\int_{-\infty}^{+\infty} d\Delta\tau \, \Delta\tau \, \exp\left[i\left(\nu\Delta\tau + \ddot{\nu}\frac{\Delta\tau^3}{3!} \right) \right] = -i2\pi \left(\sqrt[3]{\frac{2}{\ddot{\nu}}} \right)^2 Ai'\left(\nu \sqrt[3]{\frac{2}{\ddot{\nu}}} \right). \tag{3C.16}$$

Using (3C.16) in (3C.13) we can represent I as

$$I = \frac{1}{\tau_B} \left\{ \Pi(\vartheta) \, e^{i\Psi_R(\vartheta)} \, 2\pi \sqrt[3]{\frac{2}{\ddot{\nu}}} Ai\left(\nu \sqrt[3]{\frac{2}{\ddot{\nu}}} \right) \right.$$

$$\left. - i\dot{\Pi}(\vartheta) \, e^{i\Psi_R(\vartheta)} \, 2\pi \left(\sqrt[3]{\frac{2}{\ddot{\nu}}} \right)^2 Ai'\left(\nu \sqrt[3]{\frac{2}{\ddot{\nu}}} \right) \right\}, \tag{3C.17}$$

where $\nu/\sqrt[3]{2/\ddot{\nu}} \leq \zeta_0 \simeq 1$. This latter expression serves to define *proximity* as alluded to following (3C.5). We evaluate $\dot{\Pi}_R(\vartheta)$ as follows:

$$\dot{\Pi}(\vartheta) = v_{\parallel}(\vartheta)\frac{\partial\Pi}{\partial s} = v_{\parallel}\frac{\partial\Pi}{\partial\psi}\frac{\partial\psi}{\partial s}(\vartheta). \tag{3C.18}$$

From (3C.18) it is apparent that for midplane or throat resonance, $\dot{\Pi}(\vartheta) = 0$ since $\psi'(\vartheta) = 0$; and furthermore, at turning point resonance, $\dot{\Pi}(\vartheta) = 0$ since $v_{\parallel}(\vartheta) = 0$. For $k_{\parallel} \neq 0$, $v_{\parallel}(\vartheta) \neq 0$ unless $v_0 = 0$; this is the distinction between turning point and tangent resonance, the former being a special case of the latter.

It is useful to note here that treating collisional effects in the manner outlined in (3C.6)–(3C.10) has the fortunate side effect of broadening the class of resonant orbits. In the limit as the magnetic field inhomogeneity vanishes, this allows the numerical recovery of analytic results generally accessible only in this limit. Details of the calculation are described in Kerbel and McCoy (1986).

The resonant diffusion tensor elements $B_{0_{ql}} \to F_{0_{ql}}$ (cf. (3.1.103)–(3.1.106)) can be written in the form

$$B_{0_{ql}} = B_0(b*b + cc), \qquad C_{0_{ql}} = C_0(b*f + cc),$$

$$F_{0_{ql}} = F_0(f*f + cc), \qquad E_{0_{ql}} = E_0(f*b + cc). \tag{3C.19}$$

Now since the relation $B_0 F_0 = C_0 E_0$ holds everywhere, outside the proximity of tangent resonance it can be shown that $B_{0_{ql}} F_{0_{ql}} = C_{0_{ql}} E_{0_{ql}}$ and the diffusion is one-dimensional asymptotically as $\Omega \tau_B \to \infty$.

It can further be shown that for $\sin^2 \theta_0 = m\Omega_0/\omega$ the diffusion is purely energy diffusion (no pitch-angle diffusion). In fact, for small velocities, or small $k_\parallel$, these two features of resonant diffusion coalesce and become the so-called turning point or banana-tip resonance, the experimental evidence for which was referred to in Hammett *et al.* (1983) and Section 3.3.2.

Appendix 3D. Wave Models

Conventional approaches to the wave physics in the RF excitation of magnetic fusion plasma employ diverse sets of simplifying assumptions in order to reduce the difficulty of the problem. The simplest wave models deal with cold uniform plasma in a uniform magnetic field. These have the advantage of allowing direct evaluation of relatively simple expressions for wave polarizations $\mathbf{E}^\pm$ and wavenumbers $k_\parallel$, $k_\perp$ (or equivalently, index of refraction) through the cold plasma dispersion relation. However, cold plasma theory is ill-suited to treat some effects which are important in the context of wave heating (e.g., mode conversion or tunneling). As a compromise, some authors (e.g., Batchelor and Goldfinger, 1982) use cold plasma theory outside critical layers surrounding resonances and cutoffs. Wave propagation is treated in the eikonal (geometrical optics) limit by ray tracing in these exterior regions, then joined with solutions interior to the critical layers which are generated by other, usually more involved, techniques. Presuming small damping rates due to finite temperature, the absorption and thus amplitude variation along ray trajectories can then be calculated.

It is often the case that the assumptions of geometrical optics fail in the neighborhood of the critical layer and physical optics or full wave theory must be employed. The method used in the critical layers often entails the use of the local dispersion relation for warm plasma in a (locally) uniform magnetic field allowing mode conversion to/from modes not supported in cold plasma (e.g., Bernstein modes (Bernstein, 1958)). This approach has been used by Swanson (1981), Colestock and Kashuba (1983), and Phillips (Perkins *et al.*, 1984).

The net plasma heating at a given point in the wave field in the ray tracing calculations is calculated by considering all rays emanating from an antenna structure at the plasma periphery, which pass through the given point in question. Usually, Maxwellian absorption rates are computed based on a thermal plasma of presumed density and temperature, and the polarization amplitudes resulting from the wave calculation. Swanson (1980) has included the Maxwellian absorption rates in the formulation of the wave calculation resulting in a fourth-order equation. He concludes that the effect of absorption on mode conversion can be significant.

Although advanced, relative to cold uniform plasma theory, the preceding techniques are inconsistent on several significant counts. In standard warm plasma wave theory, the particle distribution function is assumed Maxwellian and the orbits are helices of invariant pitch. Therefore, any effects due to the quasilinear development of the distribution or to nonlocal wave–particle resonance phenomena are neglected. For example, particles experiencing wave–particle gyro-resonance near gyro-center turning points in a magnetic well interact particularly strongly with waves, absorbing or emitting plasma wave energy. This has the effect of distorting the distribution functions from Maxwellian form. Representing this feature, intrinsic to systems with magnetic wells in the locally uniform magnetic field theory, is awkward. A dispersion relation based on the unperturbed orbits of the system considered in the present study would be necessary to remove this limitation. Moreover, in cases of interest for ICRH studies, the eikonal treatment is only marginally valid, and for this reason, various full-wave models have been formulated (e.g., Colestock and Kashuba, 1983; Perkins et al., 1984; Swanson, 1980). In this case the fusion device is viewed as a resonant cavity containing a complex and lossy dielectric (the plasma) and a definite mode structure driven by the exciting antennae.

Including these latter refinements in the theory of RF heating in fusion devices further implies consideration of the natural response of the plasma as well as the driven response. Thus the wave field present is a combination of the wave normal modes supported by the distribution of particles present, itself evolving (slowly) self-consistently through wave–particle interaction at the frequencies corresponding to both the driving fields and the natural frequencies of the plasma. The latter arise through emission of waves due to the excitation of high frequency quasilinear currents driven by the wave–particle interaction. The analysis of successively higher generations of interacting waves is higher order in the wave amplitude and characteristic of strong turbulence.

There are three standard wave models included in the family of codes CQL. These are intended primarily as a driver package. As indicated previously, the variety of suitable wave models is quite diverse. Furthermore, many models are application specific, focusing the attention of the analysis on a small subset of related phenomena. We have therefore constructed an archetype with the required computational interconnects to the main quasilinear model. The approach here is to allow the user the freedom to adapt the code readily to his own problem with a modicum of difficulty.

Control of the main quasilinear package is mediated through a set of doubly dimensioned arrays, each containing a parameter relevant to the calculation. The first dimension of these arrays refers to the species whose excitation is under consideration, the second index refers to the mode of excitation. Since any of the RF parameters stored in these arrays can vary (over the second index) the mode structure may be as complicated as storage allows. The parameters involved are $k_\parallel$, $k_\perp$, parallel and perpendicular wave numbers; m, cyclotron harmonic number; $\mathbf{E_k}$ spectral field amplitude; $\alpha_\pm$, and $\alpha_\parallel$, the fraction of the wave field energy density in transverse and parallel polarizations; $m\Omega_0/\omega$, a normalized frequency of the RF excitation.

The doubly dimensioned array LQMODE contains literal string names denoting the particular wave model to be considered for a given (species, mode). The choices include **Icrft**, **Icrft0**, or default. The default is a one-wave point model calculation. Each wave parameter is assumed to pertain to a single wave of constant properties everywhere in the system. Of course, any number of such waves may be treated separately by choosing the array elements indexed by (species, mode) accordingly. Nondefault models establish relationships among the wave field parameters through dispersion relations. In addition, spacial, spectral, and temporal information germane to the calculation of the quasilinear resonant diffusion tensor elements is provided through the particular model specified in LQMODE (species, mode).

3D.1. Model Icrft

Model **Icrft** is a primitive one-dimensional ray tracing model for ion cyclotron (range of frequency) waves in toroidal plasmas. It uses the cold plasma dispersion relation to compute relative polarization amplitudes and perpendicular wavenumber for a given $k_\parallel$. The $k_\parallel$ spectral amplitude is calculated using the constancy of toroidal wave modenumber

$$k_\parallel R = \text{constant.} \tag{3D.1}$$

A group of waves each with a local value of $k_\parallel$ within the expected spectral range is chosen *a priori*. The initial-$k_\parallel$ spectral amplitude is specified at the plasma periphery through the (input) wave field parameters. The amplitude of each component of the local-$k_\parallel$ spectrum is then determined through (3D.1) with no attenuation (weak absorption limit). The contribution to the resonant diffusion tensor at the point in constant-of-motion space corresponding to the particle (orbit) is then determined by evaluating the trajectory integral through the local doppler shifted resonance for each wave (see Section 3.1.2). A sum over the interactions with each member of the group of waves is performed for each particle (orbit) considered.

The cold plasma theory of ion cyclotron waves in multispecies plasma is given by Stix (1962). We represent it here for completeness in the description of model **Icrft**.

Beginning with the electrical current density

$$\mathbf{j} = \sum_\alpha q_\alpha n_\alpha \mathbf{v}_\alpha \qquad (3D.2)$$

and the equation of motion for species α

$$m_\alpha \frac{d\mathbf{v}_\alpha}{dt} = q_\alpha \left(\mathbf{E} + \frac{\mathbf{v}_\alpha}{c} \times \mathbf{B} \right), \qquad (3D.3)$$

the cold plasma wave equation can be written as

$$\mathbf{n} \times (\mathbf{n} \times \mathbf{E_k}) + \mathsf{K} \cdot \mathbf{E_k} = 0, \qquad (3D.4)$$

where the dynamical variables have been represented by the plane wave ansatz $\exp[i(\mathbf{k} \cdot \mathbf{x} - \omega t)]$ and the Maxwell's equations

$$\nabla \times \mathbf{E} = -\frac{1}{c} \frac{\partial \mathbf{B}}{\partial t}, \qquad \nabla \times \mathbf{B} = \frac{4\pi}{c} \mathbf{j} + \frac{1}{c} \frac{\partial \mathbf{E}}{\partial t} \qquad (3D.5)$$

have been used.

The refractive index, or the normal slowness vector $\mathbf{n}$ used in (3D.4) is defined as $\mathbf{n} = \mathbf{k}c/\omega$ and the wavevector $\mathbf{k}$ is represented as

$$\mathbf{k} = k_\parallel \hat{\mathbf{b}} + k_\perp (\hat{\mathbf{e}}_x \cos \delta + \hat{\mathbf{e}}_y \sin \delta). \qquad (3D.6)$$

With the further abbreviations

$$P = 1 - \sum_\alpha \frac{\omega_{p\alpha}^2}{\omega^2},$$

$$R = 1 - \sum_\alpha \frac{\omega_{p\alpha}^2}{\omega^2} \frac{\omega}{\omega + \mathrm{sgn}(q_\alpha)\Omega_\alpha}, \qquad L = 1 - \sum_\alpha \frac{\omega_{p\alpha}^2}{\omega^2} \frac{\omega}{\omega - \mathrm{sgn}(q_\alpha)\Omega_\alpha}, \qquad (3D.7)$$

$$S = \frac{(R + L)}{2}, \qquad D = \frac{(R - L)}{2}.$$

Equation (3D.4) can be written as

$$\mathsf{D_k} \cdot \mathbf{E_k} = 0, \qquad (3D.8)$$

with

$$\mathsf{D_k} = \left(\begin{pmatrix} S - n_\parallel^2 - n_\perp^2 \sin^2 \delta & -(iD + n_\perp^2 \sin \delta \cos \delta) & n_\parallel n_\perp \cos \delta \\ iD + n_\perp^2 \sin \delta \cos \delta & S - n_\parallel^2 - n_\perp^2 \cos^2 \delta & n_\parallel n_\perp \sin \delta \\ n_\parallel n_\perp \cos \delta & n_\parallel n_\perp \sin \delta & P - n_\perp^2 \end{pmatrix} \right). \qquad (3D.9)$$

Now choosing $\delta = 0$ and setting $\hat{\mathbf{e}}_y \cdot \mathsf{D_k} \cdot \mathbf{E_k} = 0$ it is found that

$$\frac{E_x}{iE_y} = \frac{S - n^2}{D}. \qquad (3D.10)$$

Defining $|E_x|^2 + |E_y|^2 = E_\perp^2$ and using the fact that $(S - n^2)/D$ is real, it follows that

$$\frac{|E_y|^2}{E_\perp^2} = \frac{1}{1 + ((S - n^2)/D)^2}. \tag{3D.11}$$

The right and left circularly polarized amplitudes of the transverse electric field are given by

$$E^\pm = E_x \pm iE_y. \tag{3D.12}$$

Using (3D.11) it is easily shown that the relative polarization amplitudes can be represented as

$$\frac{|E^+|^2}{E_\perp^2} = \frac{(n^2 - R)^2}{(n^2 - R)^2 + (n^2 - L)^2},$$

$$\frac{|E^-|^2}{E_\perp^2} = \frac{(n^2 - L)^2}{(n^2 - R)^2 + (n^2 - L)^2}. \tag{3D.13}$$

Returning to (3D.8), it is apparent that a nontrivial solution $\mathbf{E_k}$ exists if and only if det $\mathbf{D} = 0$, which for $\delta = 0$ takes the form of the condition

$$n^4(P \cos^2 \theta + S \sin^2 \theta) - n^2(PS(1 + \cos^2 \theta) + RL \sin^2 \theta) + PRL = 0, \tag{3D.14}$$

where $n_\parallel = n \cos \theta$ and $n_\perp = n \sin \theta$. The reatios P/R and P/L can be approximated by

$$\frac{P}{R} \sim \frac{P}{L} \sim \frac{1 - \gamma\mu}{1 + \gamma\Omega_i/(\Omega_i \pm \omega)}, \tag{3D.15}$$

where since $\mu = m_i/m_e \gg 1$ is the ratio of ion to electron mass and $\gamma = c^2/\mu_a^2 \gg 1$ is the ratio of light to Alfven speed, $v_a^2 = B^2/mn$, (3D.15) can be approximated to $O(1/\mu)$ by

$$\frac{P}{R} \sim \frac{P}{L} \sim \mu\frac{\Omega_i \pm \omega}{\Omega_i} \gg 1 \tag{3D.16}$$

except very close to the ion cyclotron resonance. Using (3D.16) to recast (3D.14) in the approximate form

$$P(n^4 \cos^2 \theta - n^2 S(1 + \cos^2 \theta) + RL) = 0, \tag{3D.17}$$

it is readily seen that for $\omega \neq \Omega_i$ the electrostatic plasma oscillations (exhibited by $P = 0$) are decoupled from the ion cyclotron modes. For $\omega \ll \Omega_i$ we expect MHD fluid modes whose phase velocity is close to the Alfven speed. Avoiding $\theta = \pi/2$ to $O(1/\mu)$ (3D.17) can be recast as

$$n_\perp^2 = \frac{(R - n_\parallel^2)(L - n_\parallel^2)}{(S - n_\parallel^2)}. \tag{3D.18}$$

Using this result in (3D.13) a short calculation then demonstrates

$$\frac{|E^+|^2}{E_\perp^2} = \frac{(n_\parallel^2 - R)^2}{(n_\parallel^2 - R)^2 + (n_\parallel^2 - L)^2}.$$ (3D.19)

Except in the neighborhood of (cold) cyclotron resonance, these are the relations used in model **Icrft**.

3D.2. Model Icrft0

Model **Icrft0** is the same as model **Icrft** except that the polarization amplitudes and perpendicular wavenumber are arbitrarily fixed and prespecified as input wave field parameters.

References

M. Abramowitz and I. Stegun (ed.), *Handbook of Mathematical Functions*, National Bureau of Standards Applied Mathematics Series, Number 55, US Government Printing Office, Washington, DC, 446 (1968).

D. B. Batchelor and R. C. Goldfinger, *Rays: A Geometrical Optics Code for EBT*, ORNL/TM-6844 (1982).

H. L. Berk, *J. Plasma Phys.*, **20**, 205 (1978).

I. B. Bernstein, *Phys. Rev.*, **109**, 10 (1958).

I. B. Bernstein and D. C. Baxter, *Phys. Fluids*, **24**, 108 (1981).

D. T. Blackfield and J. E. Scharer, *Nucl. Fusion*, **22**, 255 (1982).

K. H. Burrell, *J. Comput. Phys.*, **27**, 98 (1978).

B. I. Cohen, R. H. Cohen, and T. D. Ronglien, *Phys. Fluids*, **26**, 808 (1983).

P. L. Colestock and R. J. Kashuba, *Nucl. Fusion*, **23**, 763 (1983).

J. W. Connor, R. C. Grimm, R. J. Hastie, and P. M. Keeping, *Nucl. Fusion*, **13**, 211 (1973).

B. Coppi and D. J. Sigmar, *Phys. Fluids*, **16**, 1174 (1973).

J. G. Cordey, K. D. Marx, M. G. McCoy, A. A. Mirin, and M. E. Rensink, *J. Comput. Phys.*, **28**, 115 (1978).

T. A. Cutler, L. D. Pearlstein, and M. E. Rensink, *Computation of the Bounce Average Code*, UCRL-52233, LLL. (1977).

H. Dreicer, *Phys. Rev.*, **115**, 23 (1959); and **117**, 329 (1960).

N. J. Fisch, PPPL-1684 and PPPL-1692 (1980).

N. J. Fisch and A. Bers, *Third Topical Meeting on RF Plasma Heating*, Caltech (1978).

N. J. Fisch and C. F. Karney, *Phys. Fluids*, **24**, 27 (1981).

R. J. Goldston, Ph.D. Thesis, Princeton University, Princeton, NJ (1977).

G. W. Hammett, J. C. Hosea, R. J. Goldston, D. Q. Hwang, R. Kaita, D. M. Manos, S. Kilpatrick, and J. R. Wilson, *25th Conference of the American Physical Society Division of Plasma Physics*, Los Angeles (1983).

R. W. Harvey, K. D. Marx, and M. G. McCoy, *Nucl. Fusion*, **21**, 153 (1981).

F. L. Hinton and R. D. Hazeltine, *Phys. Fluids*, **48**, 239 (1976).

B. H. Hui, N. K. Winsor, and B. Coppi, *Phys. Fluids*, **20**, 1275 (1977).

D. Q. Hwang, C. F. Karney, J. C. Hosea, J. M. Hovey, C. E. Singer, and J. R. Wilson, PPPL 1990 (1983).

R. Kaita, G. J. Goldston, P. Beiersdorfer, D. L. Herndon, J. Hosea, D. Q. Hwang, F. Jobes, D. D. Meyerhofer, and J. R. Wilson, *Nucl. Fusion*, **23**, 1089 (1983).

A. N. Kaufman, *Phys. Fluids*, **15**, 1063 (1972).

C. F. Kennel and F. Englemann, *Phys. Fluids*, **9**, 2377 (1966).

G. D. Kerbel and M. G. McCoy, *Comput. Phys. Commun.*, to be published, May 1986.

J. Kesner, *Nucl. Fusion*, **18**, 781 (1978).

M. Kruskal, *J. Math. Phys.*, **3**, 806 (1962).

Y. Matsuda and J. J. Stewart. *A Relativistic Multiregion Bounce-Averaged Fokker–Planck Code For Mirror Plasmas*, UCRL-92313, (1985).

M. E. Mauel, Ph.D. Thesis, MIT, Cambridge, MA (1982); also PFC/RR-82-29 (1982).

F. W. Perkins, E. J. Valeo, D. C. Eder, D. Q. Hwang, A. Kritz, C. K. Phillips, Y. C. Sun, D. G. Swanson, G. D. Kerbel, M. G. McCoy, J. Killeen, R. W. Harvey, S. C. Chiu, K. Hizanidis, V. Krapchev, D. Hewett, and A. Bers, Plasma Physics and Controlled Nuclear Fusion Research, 1984, **1**, 513 (1985).

R. D. Richtmyer and K. W. Morton, *Difference Methods for Initial-Value Problems*, 2nd ed., Interscience, New York, 1967.

M. M. Rosenbluth, R. D. Hazeltine, and F. L. Hinton, *Phys. Fluids*, **15**, 116 (1972).

M. N. Rosenbluth, W. M. MacDonald, and D. L. Judd, *Phys. Rev.*, **107**, 1 (1957).

J. E. Scharer, J. B. Beyer, D. T. Blackfield, and T. K. Mau, *Nucl. Fusion*, **19**, 1171 (1979).

L. Spitzer, *Physics of Fully Ionized Gases*, Interscience, New York, 1967.

L. Spitzer and R. Härm, *Phys. Rev.*, **89**, 977 (1953).

T. H. Stix, *The Theory of Plasma Waves*, McGraw-Hill, New York, 1962.

T. H. Stix, *Nucl. Fusion*, **15**, 737 (1975).

R. E. Stockdale, private communication (1983).

D. G. Swanson, *Nucl. Fusion*, **23**, 949 (1980).

D. G. Swanson, *Phys. Fluids*, **24**, 2038 (1981).

CHAPTER 4

A Fokker–Planck/Transport Model for Neutral Beam-Driven Tokamaks

This chapter deals with the transport simulation of neutral beam-driven tokamaks using a model which describes the background plasma by a set of fluid equations, and the energetic ions by means of Fokker–Planck equations. Section 4.1 describes this Fokker–Planck/Transport Code. Section 4.2 discusses applications to the Princeton large torus (PLT), the tokamak fusion test reactor (TFTR) and the divertor injection tokamak experiment (DITE). It should be noted that the discussion of applications to TFTR was compiled prior to the actual running of the experiment and is based on the design parameters. Comparison with the present experimental data is underway and preliminary results may be found in the work of Mirin *et al.* (1985).

4.1. Mathematical Model and Numerical Methods

Neutral beam-heated tokamaks are characterized by the presence of one or more energetic ion species which are quite non-Maxwellian, along with a warm Maxwellian bulk plasma. This background plasma may be described by a set of fluid equations. However, for scenarios in which there is a large energetic ion population, it is very important to represent the energetic species by means of velocity space distribution functions and to follow their evolution in time by integrating the Fokker–Planck equations. It is essential to utilize the full nonlinear Fokker–Planck operator to assure that the slowing-down and scattering of these energetic species are computed accurately and realistically.

The model presented here, in addition to solving one-dimensional radial transport equations for the bulk plasma densities and temperatures, solves nonlinear Fokker–Planck equations in two-dimensional velocity space for the energetic ion distribution functions. Moreover, neutral beam deposition and neutral transport are modeled using appended Monte Carlo codes developed elsewhere (Lister *et al.*, 1976; Hughes and Post, 1978).

4.1.1. Energetic ions

An arbitrary number of energetic ion species are considered, whose presence derives from the ionization and charge exchange of injected fast neutrals.

These species are described by distribution functions $f_b(v, \theta, r, t)$ in three-dimensional phase space, where b denotes the particle species, v is the velocity magnitude, θ is the pitch-angle with respect to the magnetic field, and r is the distance from the magnetic axis. It is assumed that the flux surfaces are concentric circular torii.

4.1.1.1. Fokker–Planck equations

The kinetic equation for the distribution function of energetic species b is

$$\frac{\partial f_b}{\partial t} = \left(\frac{\partial f_b}{\partial t}\right)_c + H_b - S_{bc} + S_{bcx} + \left(\frac{\partial f_b}{\partial t}\right)_E + \left(\frac{\partial f_b}{\partial t}\right)_r - L_b^\alpha - L_b^{\mathrm{orb}}. \quad (4.1.1)$$

The collision term $(\partial f_b/\partial t)_c$ is given by the complete nonlinear Fokker–Planck operator of Section 2.1.1.

It may be expressed in the form

$$\left(\frac{\partial f_b}{\partial t}\right)_c = \frac{1}{v^2} \frac{\partial}{\partial v}\left(A_b f_b + B_b \frac{\partial f_b}{\partial v} + C_b \frac{\partial f_b}{\partial \theta}\right)$$
$$+ \frac{1}{v^2 \sin \theta} \frac{\partial}{\partial \theta}\left(D_b f_b + E_b \frac{\partial f_b}{\partial v} + F_b \frac{\partial f_b}{\partial \theta}\right), \quad (4.1.2)$$

where the coefficients A_b through F_b are sums of moments of the distribution functions of all charged species present. The quantity H_b is the source resulting from the injection of neutral species b. The quantity S_{bc} represents the deceleration of energetic ions into the bulk plasma. The term S_{bcx} represents charge exchange between ion species b and the various neutral species. The quantity $(\partial f_b/\partial t)_E$ models the effect of the toroidal electric field. The term $(\partial f_b/\partial t)_r$ represents radial diffusion of the energetic ions. The quantity L_b^α is a fusion depletion term, and L_b^{orb} represents orbit losses. These terms are thoroughly described in the following sections.

The numerical solution of this type of equation has already been discussed (see Section 2.4.1). Either implicit operator splitting or the Peaceman–Rachford alternating-direction implicit (ADI) method is employed.

It is not actually necessary to solve for distribution functions f_b on every flux surface where the bulk plasma ions are defined. Treating the energetic ions in detail on every fifth flux surface combined with cubic splines of velocity-space-integrated quantities yields accurate answers in a good deal less computer time.

4.1.1.2. Neutral beam deposition

The energetic ion source term H_b is calculated using the FREYA neutral beam deposition code (Lister et al., 1976). This is a Monte Carlo code which takes into account the geometry of the tokamak and the precise locations and optical properties of the neutral beam injectors. A pseudocollision technique is employed; i.e., particle penetration is based on the minimum mean free path throughout the plasma, and resulting collisions are analyzed, *a posteriori*, to see if they are genuine or false. This pseudocollision technique enables one to compute potential collision points without calculating the intersection of the neutral path with each flux surface.

For use in the Fokker–Planck/Transport Code, several improvements have been made to FREYA:

(a) A multispecies background is allowed. That is, the neutral mean free path is based on charge exchange and impact ionization with an arbitrary number of ion species (in addition to electron impact ionization). The ionization and charge-exchange cross sections are taken from the publication of Freeman and Jones (1974).
(b) The reaction rate $\langle \sigma v \rangle$ for charge exchange and ion impact ionization is computed by averaging the product of the cross section σ and the relative velocity over the ion distribution function. A two-dimensional table look-up procedure is used.
(c) All collisions with multiply charged ions are treated as ionizations, and only one charge state of any given impurity is considered. The total reaction rate between a neutral and an impurity ion of charge Z is taken as the equivalent proton rate times $Z^{1.35}$ (Olson and Salop, 1977).
(d) When a neutral beam atom undergoes a charge exchange, its location and energy are stored for later use in the neutral transport module, enabling the modeling of multiple charge exchanges and/or reionization.
(e) The initial orbit of each deposited ion is analyzed. If that orbit strikes the limiter, the ion is discarded. This calculation assumes conservation of the toroidal component of the canonical angular momentum (Shumaker, 1979).

It is not necessary to call FREYA each timestep, as the neutral beam deposition term is usually slowly changing.

4.1.1.3. *Energetic ion deceleration*

Each energetic ion species "b" has a corresponding background plasma component. As an energetic ion decelerates, if it is not lost, it will eventually join the bulk plasma. This process is simulated by transferring all "hot" ions below a specified energy from the energetic ion distribution function to the corresponding bulk plasma component. This loss term, denoted S_{bc}, satisfies

$$\frac{\dfrac{m}{2} \displaystyle\int S_{bc}(v, \theta, r) v^2 \, d\mathbf{v}}{\displaystyle\int S_{bc}(v, \theta, r) \, d\mathbf{v}} = E_{bc}, \tag{4.1.3}$$

where E_{bc} is the average energy of transferred ions (often taken to be three halves of the electron temperature).

4.1.1.4. *Charge exchange*

The charge-exchange loss term is of the form

$$S_{bcx} = -f_b \sum_c \tilde{n}_c \langle \sigma v \rangle_{cx}^{cb}, \qquad (4.1.4)$$

where c runs over all neutral species (including neutral beam atoms) and $\tilde{n}_c$ is the corresponding neutral density. The charge-exchange rate is taken from the publication of Olson and Salop (1977). As can be seen, the charge-exchange probability is assumed to be independent of ion energy.

4.1.1.5. *Toroidal electric field*

The acceleration by the electric field in the toroidal direction is given by

$$\left(\frac{\partial f_b}{\partial t}\right)_E = -a_\parallel \frac{\partial f}{\partial v_\parallel} = -\frac{Z_b e E_\parallel}{m_b}\left(\cos\theta \frac{\partial f_b}{\partial v} - \frac{\sin\theta}{v}\frac{\partial f_b}{\partial \theta}\right). \qquad (4.1.5)$$

4.1.1.6. *Radial diffusion*

In a neutral beam-heated plasma, the fast ions will have a velocity only two to three times greater than that of the bulk ions. Thus, it is reasonable to expect that the fast ions are subject to a certain amount of radial diffusion. This is approximated by the term

$$\left(\frac{\partial f_b}{\partial t}\right)_r = -\frac{f_b}{n_b}\cdot\frac{1}{r}\frac{\partial}{\partial r}\left(rD_b \frac{\partial n_b}{\partial r}\right), \qquad (4.1.6)$$

where n_b is the hot ion density and D_b is a diffusion coefficient. This operator diffuses density but preserves velocity space shape.

4.1.1.7. *Fusion depletion*

For D–T plasmas a fusion loss term is included

$$\begin{aligned} L_D^\alpha &= \hat{n}_T \langle \sigma v \rangle_{DT} f_D, \\ L_T^\alpha &= \hat{n}_D \langle \sigma v \rangle_{DT} f_T. \end{aligned} \qquad (4.1.7)$$

Here, $\hat{n}_D$ and $\hat{n}_T$ represent the total (bulk + hot) deuteron and triton densities, and the fusion rate, which is based on a cross section given in detail by Futch et al. (1972), is taken to be independent of energy.

4.1.1.8. *Orbit losses*

Particle phase space orbits through the various meshpoints (v, θ, r) are evaluated by analyzing the toroidal component of the canonical angular momentum. This is complicated by the fact that whether or not an orbit intersects the limiter depends on the poloidal angle. It is assumed that the energetic ions are distributed uniformly with respect to poloidal angle, and an appropriate number are thrown out, based on the fraction of orbits which do interest the limiter.

4.1.2. Bulk plasma ions and electrons

An arbitrary number of bulk plasma ion species which are assumed to be Maxwellian in velocity space are considered. These species are described by densities $n_a(r, t)$ and by a common temperature profile $T_i(r, t)$. The electrons have a separately computed temperature profile $T_e(r, t)$, and their density is determined by quasineutrality; that is,

$$n_e = \underset{\substack{\text{bulk} \\ \text{plasma}}}{\sum} Z_a n_a + \underset{\substack{\text{energetic} \\ \text{ions}}}{\sum} Z_b n_b. \qquad (4.1.8)$$

4.1.2.1. Transport equations

The ion densities and the ion and electron temperatures are described by the following set of equations:

$$\frac{\partial n_a}{\partial t} = -\frac{1}{r}\frac{\partial}{\partial r}(r\Gamma_a) + \int S_{bc}\, dv\, \delta_{ab} + S_{ai} + S_{acx} - L_a^\alpha, \qquad (4.1.9)$$

$$
\begin{aligned}
\frac{\partial}{\partial t}\left(\frac{3}{2}\sum_a n_a T_i\right) = {}&-\frac{1}{r}\frac{\partial}{\partial r}\left(r\sum_a Q_a\right) + \sum_{a,b}\int S_{bc} E_{bc}\, dv\, \delta_{ab} \\
&+ \sum_a S_{ai}\tilde{E}_a + W_{cx} - \tfrac{3}{2}T_i\sum_a L_a^\alpha \\
&+ \sum_{a,b} Q_{ab} + Q_\Delta + \sum_a Q_{a\alpha},
\end{aligned}
\qquad (4.1.10)
$$

$$\frac{\partial}{\partial t}(\tfrac{3}{2}n_e T_e) = -\frac{1}{r}\frac{\partial}{\partial r}(rQ_e) + Q_{eb} - Q_\Delta + Q_{e\alpha} - \frac{3}{2}\frac{n_e T_e}{\tau_r} + j_\phi E_\phi. \qquad (4.1.11)$$

The quantities Γ_a, Q_a, and Q_e are particle and energy fluxes; E_{bc} is the mean energy of decelerated energetic ions; S_{ai} is the ionization source and $\tilde{E}_a$ is the energy of neutral species "a"; S_{acx} and W_{cx} describe charge exchange; L_a^α represents fusion depletion; Q_{ab} models heating by the energetic species; Q_Δ is energy exchange between bulk ions and electrons; $Q_{a\alpha}$ is alpha-particle heating; τ_r is the radiation loss time; and $j_\phi E_\phi$ represents ohmic heating.

4.1.2.2. Transport models

The particle and energy fluxes are written as linear combinations of the density and temperature gradients and of the toroidal electric field. This makes possible the representation of a full multispecies neoclassical transport model, as described by Mirin et al. (1977). However, present-day tokamaks do not seem to obey neoclassical scaling laws (Furth, 1975); hence, the following anomalous transport model is employed.

The particle flux Γ_a is written as

$$\Gamma_a = D_a \frac{\partial n_a}{\partial r} - R_a E_\phi, \tag{4.1.12}$$

where

$$D_a = D_{0a} + D_{1a} r^3 + \frac{D_{2a}}{n_e} + D_{3a} n_e \tag{4.1.13}$$

and

$$R_a = 2.48c \left(\frac{r}{R}\right)^{1/2} \frac{n_a}{B_\theta}. \tag{4.1.14}$$

The first term in (4.1.12) represents anomalous transport and the second term the effects of the Ware pinch.

The energy fluxes are written in terms of their convective and conductive components

$$Q_a = \tfrac{5}{2} \Gamma_a T_i + K_{ia} n_a \frac{\partial T_i}{\partial r}, \tag{4.1.15}$$

$$Q_e = \tfrac{5}{2} \Gamma_e T_e + K_e n_e \frac{\partial T_e}{\partial r}, \tag{4.1.16}$$

where

$$\Gamma_e = \sum Z_a \left(D_a \frac{\partial n_a}{\partial r} - 0.8 \, R_a E_\phi \right). \tag{4.1.17}$$

For the ion thermal conductivity the neoclassical formula of Conner (1973) is employed

$$K_{ia} = \frac{1.48c^2 (r/R)^{1/2} T_i}{e^2 B_\theta^2} \frac{m_a}{Z_a^2} \cdot \left[\frac{\langle x_a^2 v_a \rangle - \langle x_a v_a \rangle^2}{\langle v_a \rangle} \right], \tag{4.1.18}$$

where the quantity in brackets is defined by Connor (1973). For the electron thermal conductivity an empirical formula is used

$$K_e = \frac{K_{eo}}{n_e} + \frac{K_{e1}}{n_e T_e}. \tag{4.1.19}$$

The current density j_ϕ is specified (usually parabolic to the three halves power), and the toroidal electric field E_ϕ is related to the current density through

$$E_\phi = \eta_s j_\phi, \tag{4.1.20}$$

where η_s is the Spitzer resistivity (Spitzer, 1956).

4.1.2.3. *Charge exchange*

The charge-exchange source for species "a" is expressed as

$$S_{acx} = \tilde{n}_a \sum_d n_d <\sigma v>_{cx}^{ad} - n_a \sum_c \tilde{n}_c <\sigma v>_{cx}^{ca}. \qquad (4.1.21)$$

Here, the first sum runs over all charged species (including energetic ones) and the second sum runs over all neutral species (including neutral beam atoms). The term $\tilde{n}_c$ represents the density of neutral species "c", and $\langle\sigma v\rangle_{cx}^{ca}$ is the charge-exchange rate between neutral species "c" and ion species "a".

The energy gained by the bulk ions due to charge exchange is

$$W_{cx} = \sum_{d,a} \tilde{n}_a n_d \langle\sigma v\rangle_{cx}^{ad} \tilde{E}_a - \sum_{c,a} \tilde{n}_c n_a \langle\sigma v\rangle_{cx}^{ca} \cdot \tfrac{3}{2} T_i, \qquad (4.1.22)$$

where "a" runs over all singly charged bulk plasma ions, "c" runs over all neutral species (including beam neutrals), and "d" runs over all ions (including energetic ions). Recall that any charge exchange between a neutral and a multiply charged ion is treated as an ionization.

4.1.2.4. *Ionization*

The ionization source for species "a" is

$$S_{ai} = \tilde{n}_a \left(n_e \langle\sigma v\rangle_{ie} + \sum_{ions} n_b \langle\sigma v\rangle_{ib} \right), \qquad (4.1.23)$$

where electron and ion impact ionization are taken into account. As just noted, charge exchanges with multiply charged ions are included in the second term. The ionization rate formulas are based on the work of Freeman and Jones (1974).

The ionization energy source is merely equal to $\sum_a S_{ai} \tilde{E}_a$, where $\tilde{E}_a$ is the energy of neutral species "a". There is a drawback in the model, in that energetic neutrals upon ionization become part of the bulk plasma. Energy is conserved, but momentum is not. The fact that this energetic tail is assumed to thermalize to a Maxwellian instantly no doubt distorts the energy transfer with electrons.

4.1.2.5. *Radiation*

Only impurity radiation is considered. The radiation loss time is written as

$$\tau_r = \frac{\tfrac{3}{2} T_e}{\sum_Z n_Z L_Z}, \qquad (4.1.24)$$

where the sum is over all impurity species. The cooling rate L_Z is expressed implicitly as

$$\log_{10} L_Z = \sum_{i=0}^{5} A_i (\log_{10} T_e)^i, \qquad (4.1.25)$$

where the coefficients A_i are enumerated by Post *et al.* (1977). An arbitrary number of impurity species may be considered. Charge-exchange recombination radiation enhancement as proposed by Hulse *et al.* (1980) may also be implemented.

4.1.2.6. *Energy transfer*

The energy transfer rate between bulk ions and electrons is

$$Q_\Delta = \sum_a \tfrac{3}{2} n_a (T_e - T_i)/\tau_{ea}, \qquad (4.1.26)$$

where τ_{ea} is the Spitzer energy-exchange time (Spitzer, 1956). The above sum runs over all bulk plasma ions.

The heating of bulk ions and electrons by energetic ions is obtained from integrating the appropriate part of the Fokker–Planck collision operator. This results in the formula

$$Q_{ab} \sim \int_0^\infty f_a(v) v^2 \, dv \cdot \left[\int_v^\infty f_b(x) x \, dx - \frac{m_a}{m_b} \frac{1}{v} \int_0^v f_b(x) x^2 \, dx \right], \qquad (4.1.27)$$

where "a" represents the bulk species, "b" is the energetic species, and $f_{a,b}$ are the respective distribution functions. Alpha-particle heating is computed in a similar manner.

4.1.2.7. *Fusion depletion*

For D–T plasmas a fusion loss term is included

$$\begin{aligned} L_{\mathrm{D}}^\alpha &= n_{\mathrm{D}} \hat{n}_{\mathrm{T}} \langle \sigma v \rangle_{\mathrm{DT}}, \\ L_{\mathrm{T}}^\alpha &= n_{\mathrm{T}} \hat{n}_{\mathrm{D}} \langle \sigma v \rangle_{\mathrm{DT}}. \end{aligned} \qquad (4.1.28)$$

Here the symbols n_{D} and n_{T} stand for the densities of the bulk deuterons and tritons, whereas the "hatted" symbols $\hat{n}_{\mathrm{D}}$ and $\hat{n}_{\mathrm{T}}$ include both bulk plasma and energetic ion contributions. The fusion rate is taken to be independent of energy.

4.1.2.8. *Energetic ion deceleration*

The δ_{ab} appearing in (4.1.9) is a symbolic way of stating that plasma species "a" and energetic species "b" must really be the same species (e.g., both deuterons) for the transfer term to take effect. The quantity E_{bc} in (4.1.10) is the energy at which particles are transferred; in most cases, $E_{bc} = \tfrac{3}{2} T_e$.

4.1.2.9. *Discretization of the transport equations*

Equations (4.1.9)—(4.1.11) may be cast in the form

$$\frac{\partial \mathbf{u}}{\partial t} = \mathscr{L}(\mathbf{u}), \qquad (4.1.29)$$

where the vector $\mathbf{u}$ consists of the bulk ion densities and the ion and electron energy densities. An implicit, iterative difference scheme is employed; that is, (4.1.29) is approximated by

$$\frac{\mathbf{u}^{n+1} - \mathbf{u}^n}{\Delta t} = \rho \mathscr{L}_d(\mathbf{u}^{n+1}) + (1 - \rho)\mathscr{L}_d(\mathbf{u}^n), \qquad (4.1.30)$$

where Δt is the time increment, $\mathbf{u}^n = \mathbf{u}(t = n\Delta t)$, $0 \le \rho \le 1$, and the spatially discretized quantity $\mathscr{L}_d(\mathbf{u}^{n+1})$, which approximates $\mathscr{L}(\mathbf{u}^{n+1})$, is linearized with coefficients depending on the latest iterate. In particular, products of derivatives are written as

$$\left(\frac{\partial f}{\partial r}\frac{\partial g}{\partial r}\right)^{n+1} \approx \frac{1}{2}\left[\left(\frac{\hat{\partial} f}{\partial r}\right)^{n+1}\left(\frac{\hat{\partial} g}{\partial r}\right)^* + \left(\frac{\hat{\partial} f}{\partial r}\right)^*\left(\frac{\hat{\partial} g}{\partial r}\right)^{n+1}\right], \qquad (4.1.31)$$

where * refers to the latest iterate and ^ denotes a central difference approximation. Products of a function and a derivative are written as

$$\left(f\frac{\partial g}{\partial r}\right)^{n+1} \approx f^*\left(\frac{\hat{\partial} g}{\partial r}\right)^{n+1}, \qquad (4.1.32)$$

and products of functions are written as

$$(fg)^{n+1} = \tfrac{1}{2}\left[f^{n+1}g^* + f^*g^{n+1}\right]. \qquad (4.1.33)$$

Second derivatives are approximated as

$$\left[\frac{\partial}{\partial r}\left(D\frac{\partial h}{\partial r}\right)\right]_j \approx \left[\left(D\frac{\partial h}{\partial r}\right)_{j+1/2} - \left(D\frac{\partial h}{\partial r}\right)_{j-1/2}\right]\Big/\Delta r, \qquad (4.1.34)$$

with

$$\left(D\frac{\partial h}{\partial r}\right)_{j+1/2} = \left(\frac{D_j + D_{j+1}}{2}\right)\left(\frac{h_{j+1} - h_j}{\Delta r}\right), \qquad (4.1.35)$$

where the subscript j indexes the radial variable.

An exception: The ion heat convection term $\frac{5}{2}\Gamma_a T_i$ uses the latest iterate for Γ_a and treats T_i implicitly, even though Γ_a contains derivatives. That is (dropping subscripts) for $\Gamma < 0$,

$$\frac{\partial}{\partial r}(\Gamma T) \approx \frac{\Gamma_{j+1/2}T_j - \Gamma_{j-1/2}T_{j-1}}{\Delta r}. \qquad (4.1.36)$$

This linearization is appropriate for present-day transport models, in which ion heat convection dominates ion heat conduction—a fact which necessitates both implicit treatment of T_i and upwind differencing (as opposed to central differencing) of the heat convection term (Roache, 1972).

The boundary conditions are rather straightforward. At the limiter, small values of n_a, T_e, and T_i are imposed. At $r = 0$ conservation boundary conditions are employed. That is, (4.1.9)–(4.1.11) are used but with flux derivatives $-(1/r)/(\partial/\partial r)(rF)$ replaced by $-2F/r$ evaluated one-half meshpoint from $r = 0$. With the proper numerical integration scheme, the total number of ions and the total ion and electron energies are properly conserved (modulo known source and loss terms). The resulting system of difference equations is block tridiagonal, and it is solved using standard methodology (Richtmyer and Morton, 1967).

4.1.3. Neutrals

An arbitrary number of monatomic neutral species described by densities $\tilde{n}_a(r, t)$ and mean energies $\tilde{E}_a(r, t)$ are considered. These neutrals result from: (1) charge exchange of injected beam neutrals; (2) gas puffing; and (3) recycling from the limiter and wall. Neutral transport is computed using the AURORA code of Hughes and Post (1978). Although AURORA is a three-dimensional Monte Carlo code, it does not take into account toroidal effects, but instead assumes a long, straight cylinder. This, of course, results in some inaccuracies in the treatment of energetic neutrals. AURORA does not use a pseudocollision technique. The local mean free path and distance traveled per zone must be computed for each particle. It is the time-consuming nature of this latter computation which necessitates the assumption of a cylindrical geometry rather than a toroidal one.

As is the case with FREYA, several improvements have been made in AURORA. It is now a multispecies neutrals transport code. An arbitrary number of charge exchanges involving an arbitrary number of species may be considered. The reaction rates $\langle \sigma v \rangle$ are computed as in FREYA, and all collisions with multiply charged ions are treated as ionizations. In addition, neutrals can be launched from any radius, thereby enabling consideration of neutrals arising from charge exchange of injected beam neutrals. The neutral density profiles computed by AURORA are scaled to yield the correct integrated ionization rate. Also, neutral transport need not be computed every timestep, as that procedure would be too time-consuming.

4.1.4. Fusion

There are three contributions to the fusion reaction rate: (i) thermonuclear reactions, denoted R_{11}; (ii) "beam–target" reactions, denoted R_{12}; and (iii) reactions among the energetic ions, denoted R_{22}. At each plasma radius, the fusion reactivities $\langle \sigma v \rangle_{11}$, $\langle \sigma v \rangle_{12}$, and $\langle \sigma v \rangle_{22}$ are evaluated numerically via a fivefold velocity space integral (Marx et al., 1976; Cordey et al., 1978):

$$R_{ij} = \int f_i(\mathbf{v}_i) f_j(\mathbf{v}_j) \sigma(\mathbf{v}_i - \mathbf{v}_j) |\mathbf{v}_i - \mathbf{v}_j| \, d\mathbf{v}_i \, d\mathbf{v}_j. \tag{4.1.37}$$

The R_{ij} are then integrated over the plasma volume, to give the total reaction rate.

4.1.4.1. Deuteron plasmas

In deuteron plasmas two types of fusion reactions occur:

$$D + D = T + p + 4.04 \text{ MeV},$$
$$D + D = {}^3H_e + n + 3.27 \text{ MeV}. \tag{4.1.38}$$

Each reaction probability is computed separately based on cross sections found in the work of Futch et al. (1972). Thus, both the neutron production rate and the total fusion power may be monitored. Because these reactions occur at such a slow rate, it is not necessary to include fusion depletion terms nor is it necessary to consider the effects of reaction products.

4.1.4.2. Deuteron–triton plasmas

Here it is necessary to consider only the reaction

$$D + T = \alpha + n + 17.58 \text{ MeV}, \tag{4.1.39}$$

as the number of D–D reactions will be orders of magnitude smaller. The fusion cross section may be found in the publication of Jassby (1977). Unlike the D–D case, the effects of the resulting fusion products (namely alpha-particles) must be considered.

The alpha-particle velocity distribution is taken to be the angle-averaged distribution given by Jassby (1977). Alpha heating is computed through integration of the Fokker–Planck collision operator. For computational convenience, all heat destined to be transferred from the alphas to the energetic ions is added to the bulk plasma ions instead. The alpha-particle density is reduced in order to take into account the fact that some of the alpha-particles will be lost on their first bounce. For this purpose the subroutine of Shumaker (1979) is employed. Depletion of deuterons and tritons as a result of fusion is also modeled. This treatment of $f_\alpha(v)$ is reasonable only when plasma temperatures are changing slowly.

4.2. Applications

4.2.1. Princeton large torus

During 1978 the Princeton large torus (PLT) achieved record-setting temperatures. At high beam powers ($P_B \approx 2.4$ MW) and low plasma densities ($n_e(0) \lesssim 5.5 \times 10^{13}$ cm^{-3}), ion temperatures as high as 6.5 keV were reported (Eubank et al., 1979). Moreover, the fractional hot density on the magnetic axis was measured to be up to 30%, and theoretical analyses indicate that at low density, the majority of the fusion neutrons resulted from either beam–beam or beam–target reactions (Colestock *et al.*, 1979).

In this chapter, results of the above model are compared with detailed experimental data obtained from PLT in August 1978. First, the predicted and measured neutron emissions are compared, using the experimentally measured plasma profiles in the code in order to isolate the behavior of the energetic ions. Next, the full computational model is utilized to evaluate and compare important measurable quantities such as neutron flux and electron temperature, and particular attention is paid to the effects of varying the assumed transport model.

4.2.1.1. *Energetic ion behavior*

The assumptions made in the code concerning the behavior of the injected energetic ions while slowing down can best be tested by using the experimentally measured profiles of electron density, n_e, electron temperature, T_e, as well as the experimental values of the spatially averaged toroidal electric field and impurity content, Z_{eff}. Here $Z_{eff} = \sum_a Z_a^2 n_a/n_e$ is assumed to be uniform in radius and due only to carbon. The profile of ion temperature, T_i, is assumed to be the same as that of T_e, while $T_i(0)$ is estimated from charge-exchange measurements. The profile of neutral density is taken as $a + be^{cr}$ ($a = 1.35 \times 10^8$ cm^{-3}, $b = 1.2 \times 10^6$ cm^{-3}, $c = 0.1776$), which is consistent with the predictions of Monte Carlo neutrals codes for plasmas of PLT size. A toroidal electric field consistent with the experimentally measured loop voltage is applied. Eighty-five percent of the neutral beam power is at the full energy and 15% is at the half energy. Some of the principal PLT parameters are listed in Table 4.1. In Table 4.2 are listed the experimental shots used for comparison purposes along with some of their characteristics. For more detailed information see the work of Colestock *et al.* (1979).

With the above input, the Fokker–Planck equations for the energetic ions are iterated to steady state. During this iteration, the bulk plasma ion density n_i is dynamically adjusted to maintain the prescribed n_e and Z_{eff}, according to

Table 4.1. PLT parameters.

Major radius	1.40 m
Minor radius	0.40 m
B-Toroidal	3.2 T
Plasma current	~0.5 MA
Neutral beam energy	~35 keV
Neutral beam power	up to 2 MW
	85% full energy
	15% half energy
Injection angles	0°, 180°

Table 4.2. PLT experimental shots.

Shot	$\bar{n}_e (\times 10^{13} \text{ cm}^{-3})$	P_{co}(MW)	$P_{counter}$(MW)	Z_{eff}	Identifying Letter
88215	2.3	0.95	—	1.6	A
88203	3.1	0.44	—	1.7	B
88204	3.7	0.97	—	3.4	C
88193	4.6	0.49	—	1.2	D
88216	2.4	—	0.85	1.6	E
88206	3.5	—	0.54	1.9	F
88205	4.1	—	0.90	2.1	G
88186	5.8	—	0.28	1.1	H
88188	6.2	—	0.29	1.1	I
88222	1.7	1.03	0.30	2.7	J
88226	2.0	0.98	0.76	2.8	K
88221	2.1	0.58	0.86	2.7	L
88220	2.1	1.00	0.80	2.3	M
88219	2.5	0.95	0.82	2.3	N
88214	3.1	0.91	0.90	1.6	O
88201	4.6	0.89	0.84	2.7	P
88198	4.8	0.52	0.33	1.3	Q
88192	5.0	0.52	0.52	1.3	R
88189	5.9	0.45	0.73	1.1	S
88182	6.2	0.45	0.25	1.1	T

$n_e = n_i + n_h + 6n_c$. The computed neutron fluxes are then compared with the experimentally measured values. This comparison is carried out using data toward the end of the beam pulse—when the neutron production rate is at its peak and when the experiment itself has reached a quasi-steady state. Results are shown in Fig. 4.1. In this and several of the other figures, each letter refers to a particular experimental shot. The correspondence is included in Table 4.2.

The agreement between code predictions and experimental results is within a factor of 2 over a wide range of beam power (0.3 to 2.3 MW) and plasma density (the line-averaged electron density $\bar{n}_e = (1/a) \int_0^a n_e(r) \, dr$ ranges from 1.5 to $6.2 \times 10^{13} \text{ cm}^{-3}$). The agreement is best for low-density, high-power cases in which there is simultaneous and co- and counter-injection. For other scenarios the code prediction generally exceeds the experimental measurement.

The principal uncertainties are as follows.

(a) Limited experimental results have been used. Comparisons have been carried out with shots of August 1, 1978. Accurate data for other dates had not been made available at the time that the numerical simulations were carried out.

(b) Accurate measurements of $T_i(r)$ do not exist. A 25% increase in T_i will add significantly to the computed beam–target and thermal ion fusion rates, resulting in a 25% increase in the neutron flux F_n. Calculations in which $T_i(r)$ is assumed to be 50% greater than the estimate causes an even larger increase in F_n, primarily through additional bulk ion fusion reactions.

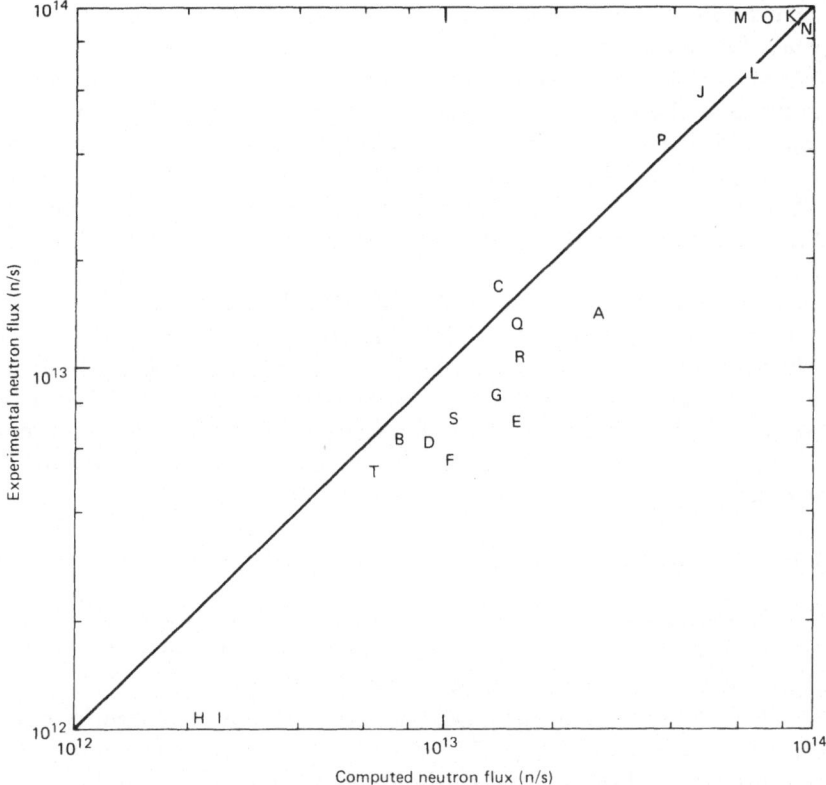

Figure 4.1. Computed neutron fluxes versus experimentally measured values for steady-state calculations with fixed bulk plasma parameters (PLT).

(c) Accurate measurements of $n_n(r)$ do not exist. Results of time-dependent simulations in the next section indicate that the neutral density on axis might be an overestimate, especially for high-density, low-injection-power cases, and this can cause a distortion in the charge-exchange loss rate. Calculations in which $n_n(0)$ is lowered to 6.9×10^6 cm^{-3} show an increase in F_n of at most 25%.

(d) The experiment is never at an exact steady state. This is not thought to be very important, since experimental measurements indicate that the neutron flux settles down well before the end of the beam pulse, at which time the plasma parameters are changing rather slowly.

(e) Z_{eff} is not independent of radius, and impurities other than carbon are present. This will have a small effect on the beam slowing-down but a possibly larger effect on beam penetration, since our model assumes collision probabilities proportional to $Z^{1.35}$.

(f) No hot ion diffusion is assumed. For high-density, low-power cases the beams do not penetrate well, and the computed hot ion density profile is peaked off-axis. If the energetic ion diffusion rate D_h is comparable to the assumed bulk ion diffusion rate D_a, both inward and outward diffusion occur, and the neutron flux does not change significantly. However for cases in which the fraction of energetic ions is significant ($\sim 10\%$ on axis), a D_h comparable to D_a results in a 10% decrease of F_n.

(g) The plasma column has a tendency to shift off-axis. Calculations in which the beams are aimed 9 cm off-axis show at most a 5% change in neutron flux.

(h) The beam current and voltage will tend to fluctuate with time. The error, of course, depends on the degree of fluctuations.

(i) The neutron flux measurement may not be calibrated correctly. Even with these uncertainties, code and experiment do agree to within a factor of 2. Moreover, the agreement is best for scenarios in which the beam–beam interactions are important, namely co- and counter-injection (~ 2 MW) at low density.

4.2.1.2. Time-dependent simulations

Having established that the code realistically models the energetic ions, time-dependent modeling of the beam injection phase of PLT is considered. Direct comparisons are again made with August 1978 PLT shots.

Initially (that is, at beam turn-on), the bulk deuteron and impurity density profiles are assumed to vary as $(1 - r^2/a^2)^3$. Carbon and iron impurities are chosen with $n_{iron} = 0.1 n_{carbon}$ and Z_{eff} uniform in radius. The magnitudes of the density profiles are determined by matching the electron line density to the experimentally measured value. The electron temperature and bulk ion temperature profiles also vary as parabolic cubed; the initial central electron temperature is taken to be the experimentally measured value at 150 ms into injection, and the initial central ion temperature is defined according to $T_i(0) = T_e(0) (0.25 + 8 \times 10^{-13} \bar{n}_e)$. This reflects the fact that prior to beam turn-on, high-density cases have a stronger coupling between T_e and T_i. It should be noted that the profiles after beam injection do not depend strongly on the values prior to beam injection. The energetic ion density is, of course, assumed to be initially zero.

At the limiter $n_e \approx 2 \times 10^{12}$ cm^{-3} and T_e and T_i equal 10 eV. These conditions correspond to those of present-day tokamak plasmas, as far as can be determined. Moreover, the overall results are insensitive to the exact edge densities and temperatures provided they are small enough.

The code is then run for 150 ms, which is the approximate duration of beam injection. The amount of gas puffing in the code is dynamically determined to match the experimentally measured electron line density. The impurity density profiles are adjusted to a uniform Z_{eff}, and a recycling coefficient (defined as the neutral influx divided by the ion outflux) of $R_c = 0.9$ is prescribed. Ninety

Table 4.3. Principal results of FPT simulations of PLT.

Shot	$F_n(\times 10^{13} s^{-1})$	$T_e(0)$ (keV)	$T_i(0)$ (keV)	τ_n(ms)[a]
88215	1.73	1.48	2.00	8.4
88203	0.31	0.99	0.99	3.6
88204	1.40	1.38	1.53	5.5
88193	0.46	0.92	0.88	2.5
88216	1.12	1.33	1.64	5.2
88206	0.56	1.03	1.02	3.3
88205	1.02	1.18	1.21	3.4
88186	0.11	0.76	0.71	1.2
88188	0.07	0.75	0.70	1.6
88222	5.69	2.03	3.86	11.7
88226	11.12	2.37	4.73	13.1
88221	7.35	2.16	3.81	11.5
88220	9.62	2.25	4.30	11.5
88219	8.82	2.24	3.81	10.5
88214	6.96	2.14	3.23	9.7
88201	4.23	2.07	2.23	6.4
88198	0.91	1.14	1.11	2.8
88192	1.24	1.22	1.22	3.1
88189	1.14	1.21	1.17	2.4
88182	0.51	0.90	0.86	1.8

[a] Neutron flux decay time.

percent of the neutrals which strike the limiter are assumed to be reflected at 40 eV; the rest are absorbed.

Our primary transport model has $D_{0a} = D_{1a} = D_{3a} = K_{e0} = 0$, $D_{2a} = 5 \times 10^{16}$, and $K_{e1} = 2.4 \times 10^{17}$. Here n_e is in cm^{-3} and T_e is in keV. Key results are listed in Table 4.3.

4.2.1.2.1. *Fusion neutron production.* Figures 4.2, 4.3, and 4.4 show comparisons between the computed neutron fluxes and the experimentally measured values at $t = 150$ ms. In Fig. 4.2 is plotted the computed flux divided by the experimental flux as a function of injection power. The results generally agree to within 30%, and some of the best results occur at around $P_B = 2$ MW, where beam–beam reactions are important. In Fig. 4.3 is plotted the same ratio, but now as a function of electron line density, $\bar{n}_e$. The agreement is best at low $\bar{n}_e$, where the role of the energetic ions is most important. In Fig. 4.4 is plotted the computed flux versus the experimental flux. There is excellent agreement as F_n varies over two orders of magnitude.

Figure 4.5 illustrates the origin of the fusion reactions for the co- plus counter- shots. In all cases the neutrons originate predominately from beam–target reactions. At low density almost as many neutrons are produced by beam–beam reactions because the longer slowing-down time for the hot ions results in a larger hot ion density. At high density virtually all of the fusion is

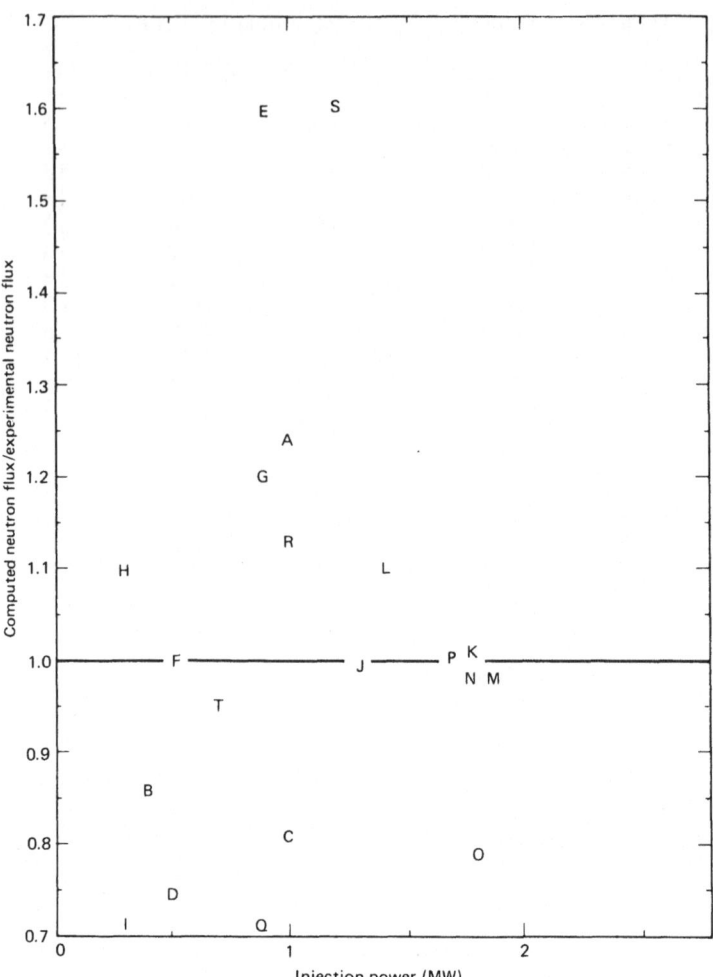

Figure 4.2. Computed neutron flux divided by experimental neutron flux versus injected beam power (PLT).

due to beam–target reactions since there are relatively few hot ions in the first place, and since the low bulk ion temperature mitigates against thermal reactions. Except at low densities, thermonuclear and beam–beam reactions are comparable. In fact, the code model slightly overestimates the number of bulk ions at the expense of the number of energetic ions.

In summary, agreement between code and experiment is best at high injection power, low density, and for cases in which there is both co- and counter-injection. Moreover, the majority of the neutrons result from beam–target reactions. As is evident from Fig. 4.5, if very low densities could be obtained by

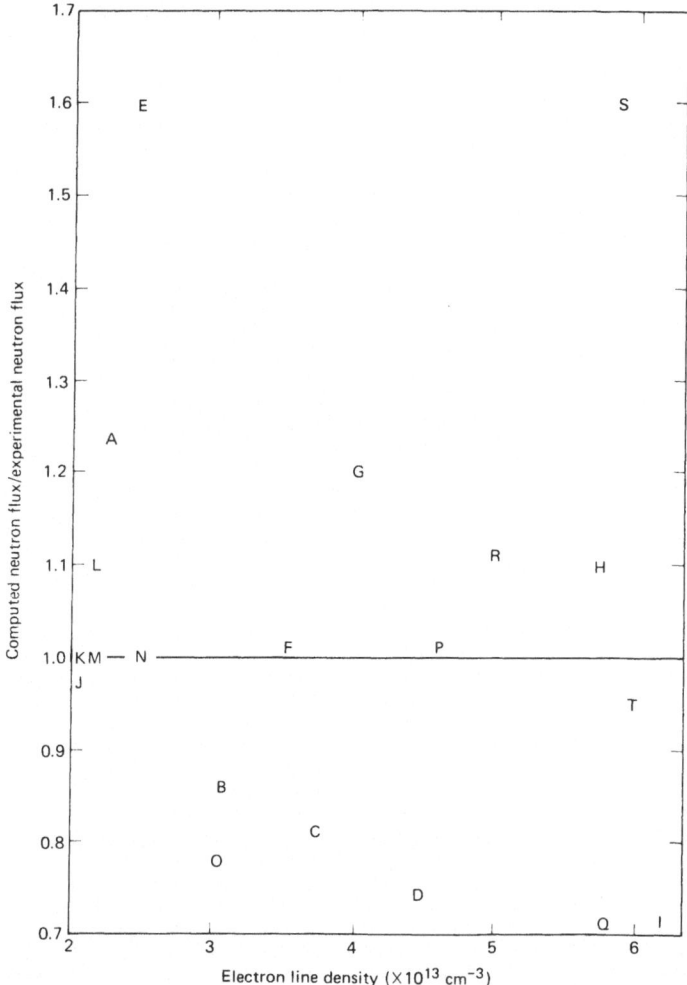

Figure 4.3. Computed neutron flux divided by experimental neutron flux versus electron line density (PLT).

reducing the recycling coefficient, then beam–beam reactions would dominate, realizing a CIT (counterstreaming-ion torus) plasma.

4.2.1.2.2. *Electron and ion temperatures.* In Fig. 4.6 the experimental values of T_e on axis are compared with the computed values. The code consistently understimates the experimental $T_e(0)$ by about 25%, even with $K_e \sim 1/T_e$.

A more detailed look at the temperature profiles is taken in Fig. 4.7(a) and (b). Here, the experimentally measured T_e, the computed T_e, and the computed T_i profiles are plotted for two of the PLT shots. Run number 88182 is a low-

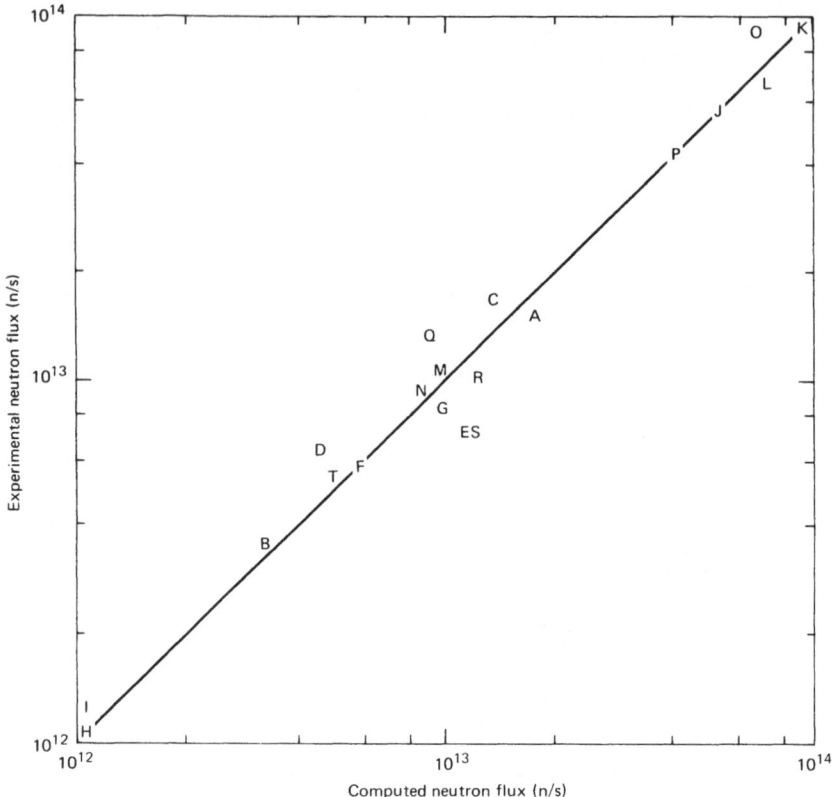

Figure 4.4. Computed neutron flux versus experimental neutron flux (PLT).

power, high-density case, and run number 88214 is a high-power, medium-to-low density case. Both have co- and counter-injection. We see that in each case, the computed T_e is consistently lower than the experimental T_e. The degree to which the lower T_e affects the neutron flux is examined by redoing the steady-state cases (previous section) for these shots. A 20% drop in the assumed T_e profile causes the neutron flux to decrease by 10–15%.

Figure 4.7(a) and (b) also contain plots of the bulk ion temperature profiles. In run 88182, the high density dictates short slowing-down and energy exchange times, resulting in $T_i \approx T_e$. In run 88214, however, T_i is 50% greater than T_e. Moreover, the ratio T_i/T_e is a strong function of radius, and this might have great bearing on the electron transport (Molvig *et al.*, 1979).

4.2.1.2.3. *Electron heat conduction.* Next the effect of varying the magnitude and shape of the electron heat conductivity is investigated. Detailed comparisons of the electron temperature profiles for runs 88182 and 88214 are

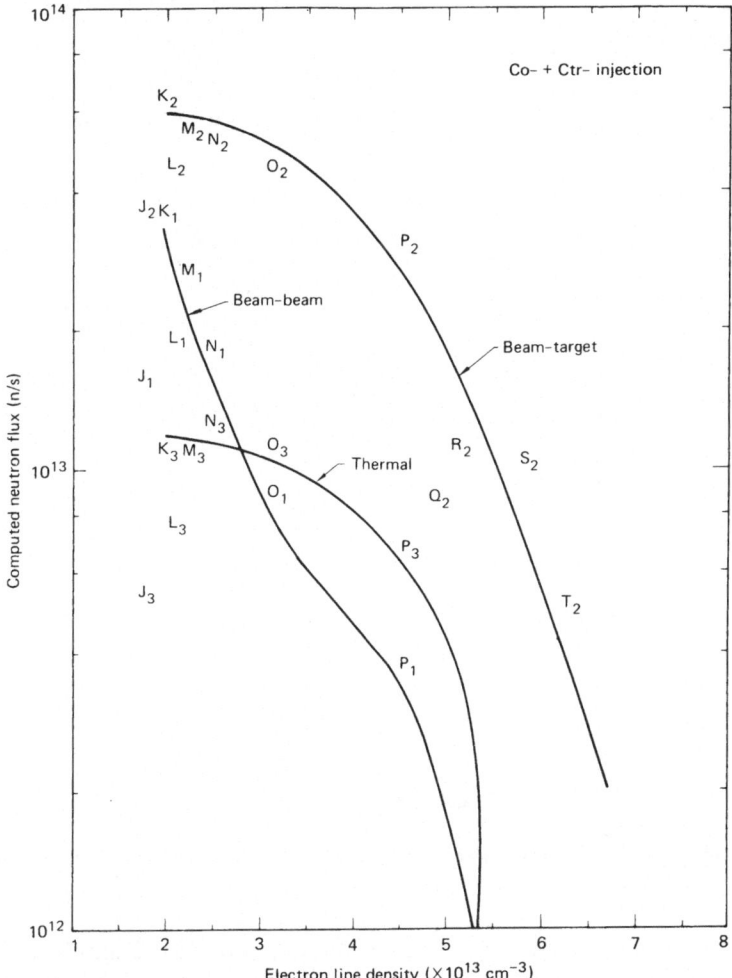

Figure 4.5. Breakdown of fusion reactions according to type (PLT).

shown in Figs. 4.8 and 4.9, respectively, for the same particle diffusion coefficients. For run 88182, the average of all the transport models seems to be a relatively good approximation, except on axis. In run 88214, the computed T_e is lower than the experimentally measured value, and the experimental profile is more peaked on axis than any of the computed profiles. The shape and magnitude of the electron temperature varies considerably as the electron thermal conductivity K_e is varied. Comparisons with other experimental shots indicate that this trend is not uncommon. Thus, it is difficult to cite a particular transport model or transport coefficient as being truly appropriate.

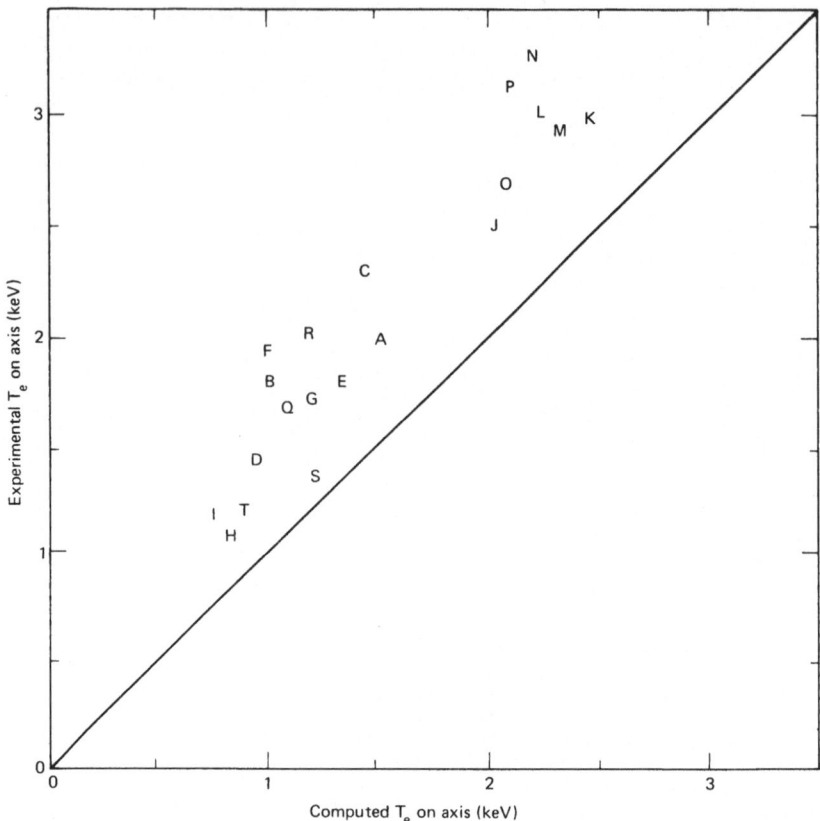

Figure 4.6. Computed electron temperature on axis versus experimental value (PLT).

4.2.1.2.4. *Diffusion of energetic ions.* Results of runs 88182 and 88214 with a varying energetic ion diffusion coefficient D_h are shown in Figs. 4.10 and 4.11, respectively. In run 88182, the high density causes poor beam penetration, and the resulting hot ion density profile is peaked off-axis. Radial diffusion of the hot ions does cause the profile to flatten somewhat, but its effect is limited because of the short slowing-down time. In run 88214, n_{hot} is peaked very close to the axis (in the absence of diffusion), and the inclusion of radial diffusion combined with a larger slowing-down time causes n_{hot} to flatten out and to peak on axis. There is a corresponding drop of about 20% in the peak neutron flux.

4.2.1.2.5. *Neutron decay.* To simulate the decay phase of the experiment, the code is run for one neutron decay time τ_n after beam turn-off. That is, at 150 ms the beam currents are set to zero and the code is run until the neutron flux diminishes from its peak value by a factor of e (2.71828…). During this decay

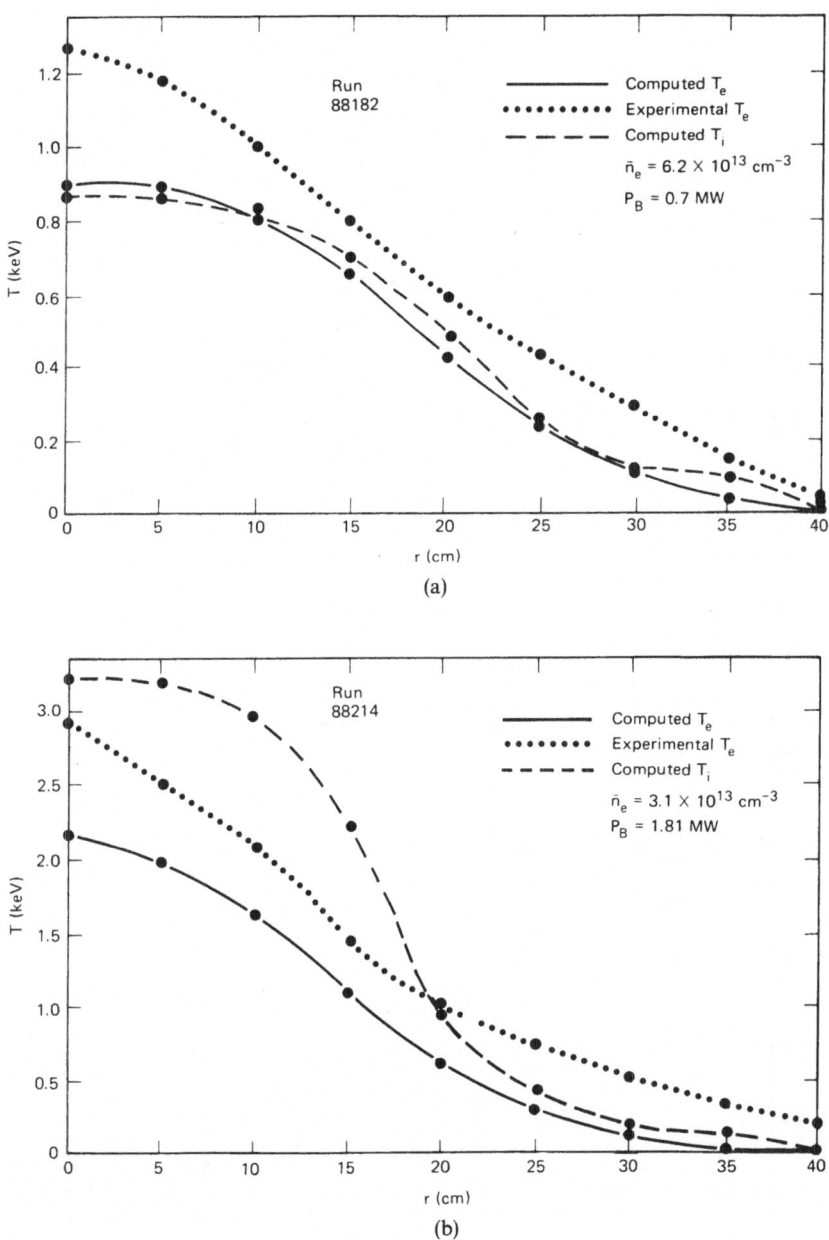

Figure 4.7. Computed electron and ion temperature and experimentally measured electron temperature profiles versus radius—PLT runs (a) 88182 and (b) 88214.

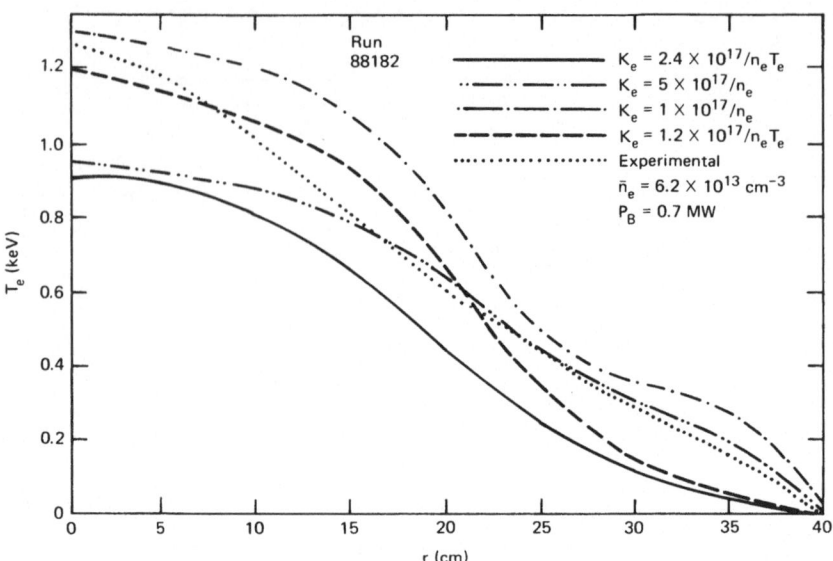

Figure 4.8. Electron temperature radial profile versus transport model—PLT run 88182.

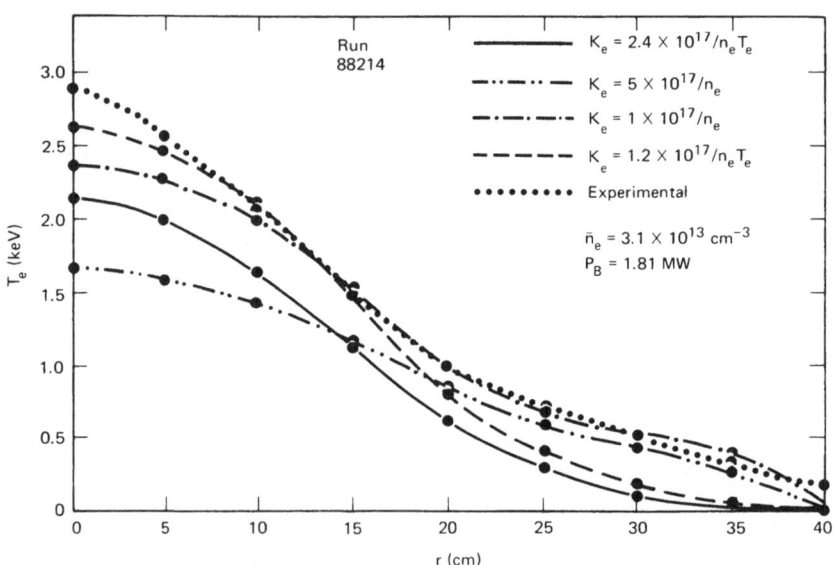

Figure 4.9. Electron temperature radial profile versus transport model—PLT run 88214.

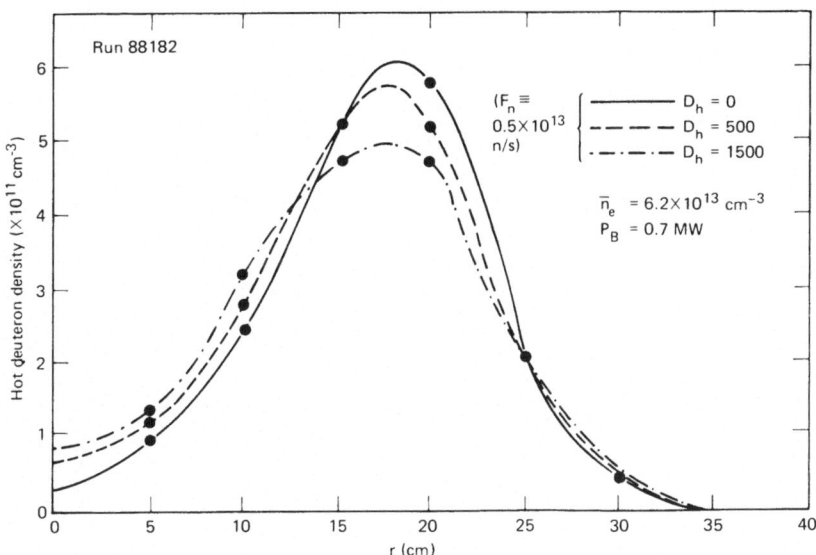

Figure 4.10. Hot deuteron density radial profile versus energetic ion diffusion—PLT run 88182; D_h is in cm^2 s^{-1}.

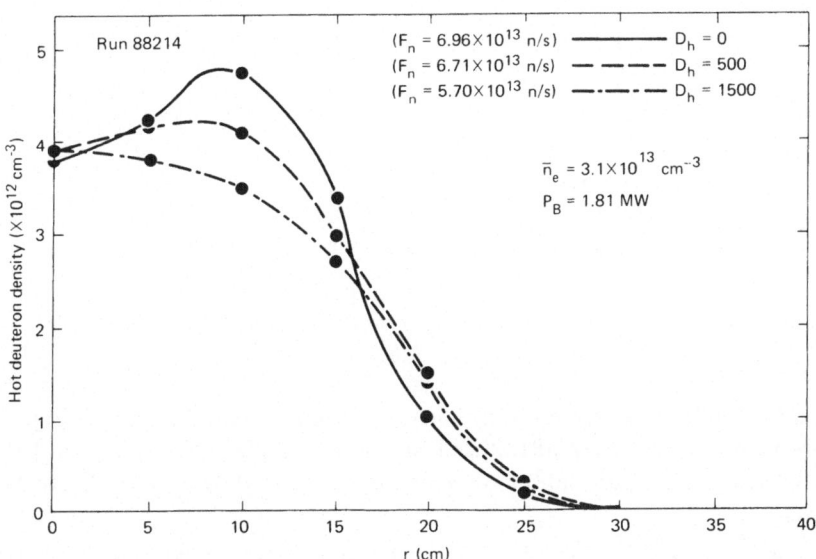

Figure 4.11. Hot deuteron density radial profile versus energetic ion diffusion—PLT run 88214; D_h is in cm^2 s^{-1}.

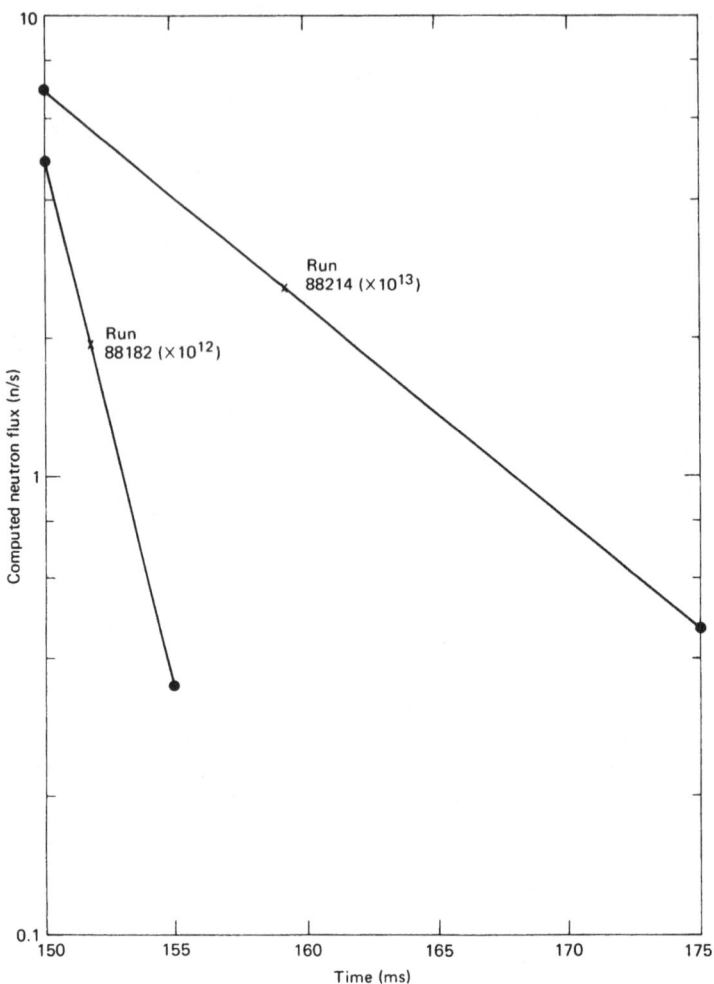

Figure 4.12. Neutron flux versus time after beam turnoff for PLT runs 88182 and 88214.

phase F_n is indeed an exponential function of time, as can be seen in Fig. 4.12, where results from shots 88182 and 88214 are displayed. There is not sufficiently accurate experimental data available with which to compare these results.

4.2.1.2.6. *Uncertainties*. In Section 4.2.1.1 it was relatively easy to list uncertainties, since there, most of the bulk plasma parameters used in the code were taken from the experimental data. Now, however, the fact that the bulk plasma is also varying with time adds a whole new dimension. As has already been demonstrated, a lack of knowledge of the transport model is a very large uncertainty. The interaction between the plasma and the wall is another area which is not well understood. The ingoing and outgoing particle fluxes are so large that a change in the recycling coefficient of 0.05 has an enormous effect on the resulting density. Given accurate time-histograms of line density and gas puffing rate, it would be possible to estimate R_c (modulo transport losses). Instead it has been necessary to assume a constant R_c of 0.9 and to dynamically vary the gas puffing rate.

Another area of uncertainty is the modeling of the impurities. To do this accurately requires knowledge of the impurity transport coefficients as well as the dynamics of the impurity-wall interactions. Lacking this knowledge, it has been expedient to assume a temporally and spatially independent impurity mix.

4.2.2. Tokamak fusion test reactor

One of the primary objectives of the TFTR (tokamak fusion test reactor) is to demonstrate fusion energy "break-even" in a deuterium–tritium plasma (Spano, 1975; Jassby, 1977). Here break-even is defined as $Q_p = 1$, where $Q_p =$ (fusion power production/injection heating power). The most straightforward means of achieving $Q_p = 1$ in a tokamak device is by the injection of energetic neutral deuterium beams into a tritium target plasma, an approach called the TCT (two-energy-component torus) (Dawson et al., 1971; Furth and Jassby, 1974). For obtaining values of Q_p significantly greater than 1, however, the target plasma should contain some optimal composition of tritium and deuterium (Jassby, 1977). In principle, the appropriate ratio of D and T in the target plasma can be maintained by the programming of gas or pellet injection. In practice, significant deuterium fueling by the injected beams and recycling of exiting plasma at the limiter and wall will make it difficult to control the target plasma composition (Jassby and Towner, 1976). One way to effect such control is to fuel the plasma by injection of both tritium and deuterium neutral beams. While this approach may not give as large a Q_p as the ideal TCT system, it will enable the plasma composition to be constant over extended periods, and thus should offer a more stable operating regime for a beam-driven reactor of moderate Q_p.

There is presently considerable interest in the possibility of reaching higher Q_p-values in the TFTR, and in particular of attaining $Q_p \geq 2$. Using either D^0 beams only, or both D^0 and T^0 beams, there are several possible approaches to this goal:

(1) Using D^0 injection only, one can rely mainly on beam–target reactions, with an optimal background composition of about 75% T and 25% D (Jassby, 1977),
(2) One can set up the so-called CIT (counterstreaming-ion torus) system (Kulsrud and Jassby, 1976), with the injection of oppositely directed D^0

and T^0 beams, and with the plasma recycling coefficient kept as low as possible (with a gettering system or unload divertor). Fusion energy is produced predominately by beam–beam and beam–target reactions.

(3) One can establish a BDTN (beam-driven thermonuclear) plasma, with most of the fusion reactions occurring in a thermal plasma of composition approximately 50 : 50 D–T. This plasma is heated and fueled by both D^0 and T^0 injection, or by D^0 beams alone.

In all three methods, it is likely that T_i will significantly exceed T_e because of beam fueling, and because more than 50% of the beam power can flow to the thermal ions (Jassby *et al.*, 1977). The BDTN plasma requires a higher $n\tau_E$, but it is capable of higher Q_p values than the TCT- or CIT-type systems. The BDTN system with D^0 and T^0 injection is actually a natural evolution of the CIT system, when both $n\tau_E$ and the plasma recycling coefficient are increased (Jassby and Towner, 1976).

Table 4.4 gives the reference machine and injector parameters for the TFTR. The purpose of the present study is to compare the performance of these methods of obtaining $Q_p > 1$ for the TFTR parameters, although the results are certainly applicable to other large beam-driven toroidal reactors. Regardless of the operating mode used to obtain $Q_p = 1$ to 2, reactions involving the energetic ions directly are an important—or dominant—component of the total reaction rate. While beam–injection experiments have revealed that the slowing down rate of fast ions is classical, relatively little is known of their radial diffusion. An important aspect of the present work is to investigate how radial diffusion of the fast ions can affect fusion performance.

The following sections examine the time-dependence of plasma parameters and their dependences on beam power, injection scenario, transport model, and recycling rate.

Table 4.4. TFTR parameters.

Major radius	2.48 m
Minor radius	0.85 m
B-Toroidal	5.2 T
Plasma current	2.5 MA
Neutral beam energy	Up to 120 keV D^0
	Up to 150 keV T^0
Neutral beam power	Up to 40 MW (total)
	80% at full energy
	20% at half energy
Tangency radius	0 (perpendicular injection)
	2.30 m (tangential injection)
Beam divergence	0.3° horizontal
	0.7° vertical
Beam pulse length	1–5 s

4.2.2.1. *Plasma parameters*

It is assumed that 80% of the neutral beam power is at the full energy and 20% at the half energy. At the limiter radius, $n_e \approx 1 \times 10^{13}$ cm^{-3} and T_e and T_i equal 100 eV. The starting parameters (at $t = 0$) are $T_e \approx 5\,(1 - r^2/a^2)$ keV, $T_i \approx 10\,(1 - r^2/a^2)$ keV, and $n_e \approx 5 \times 10^{13}\,(1 - 0.8r^2/a^2)$ cm^{-3}. These initial temperatures are much higher than can be attained by ohmic heating, but are used to shorten the computational time. Calculations starting with realistic ohmic heating conditions show that the above temperatures are attained within about 150 ms of beam injection at 24 MW.

As a representative impurity, iron is used with a density of 3×10^{10} cm^{-3}, independent of radius and time. Since there is little definitive experimental information concerning the spatial distribution of impurities under arbitrary conditions, it would be fruitless to include a more detailed model. Other likely impurity ions in the TFTR plasma are oxygen and carbon (from the limiter). The inclusion of impurity radiation from these ions would change the power balance conditions only near the plasma boundary. Our code results show that the temperature and density characteristics of the central plasma region are relatively insensitive to the temperature near the plasma edge. The principal effect of low-Z impurity concentration would be depletion of the reacting ion population. If carbon and oxygen can be largely removed by the planned gettering systems in TFTR, then these atoms will have little effect on the fusion performance. Wall sputtering, which would increase the ion content with time, is excluded from our analysis.

The instantaneous fusion power multiplication Q_p is

$$Q_p = \frac{\int \sum R_{ij}\, \vec{dr} \cdot E_f}{\text{injected beam power}}, \tag{4.2.1}$$

where E_f is the fusion energy produced per reaction and R_{ij} is given by (4.1.37). The fusion alpha power is $0.20 \times Q_p \times P_{\text{beam}}$, where P_{beam} is the injected beam power. Separate code calculations show that over 90% of the alphas born on the magnetic axis will be confined for at least one bounce. Taking into account the spatial variation in alpha production, the progressively poorer first-bounce confinement at larger radii, and the fact that some alphas will be lost on later bounces, the average confinement fraction varies from 60% to 80%. It is assumed here that just two-thirds of the alphas are confined.

The alpha-particle distribution function is computed as described in Section 4.1.4.2. This treatment of $f_\alpha(v)$ is reasonable only when plasma temperatures are changing rather slowly; in the present study, this assumption corresponds to times $t \gtrsim 1.5$ s (see Section 4.2.2.2). The values of Q_p quoted in this chapter always refer to this quasisteady period. Radial diffusion of the fusion alpha-particles is not a relevant phenomenon for the TFTR plasma, because of the large orbit size of the alphas. In fact, the widths of the largest banana orbits of the fusion alphas are comparable with the plasma radius, for $I_p = 2.0$ to 2.5 MA. Thus orbit confinement is the domiant spatial loss mechanism.

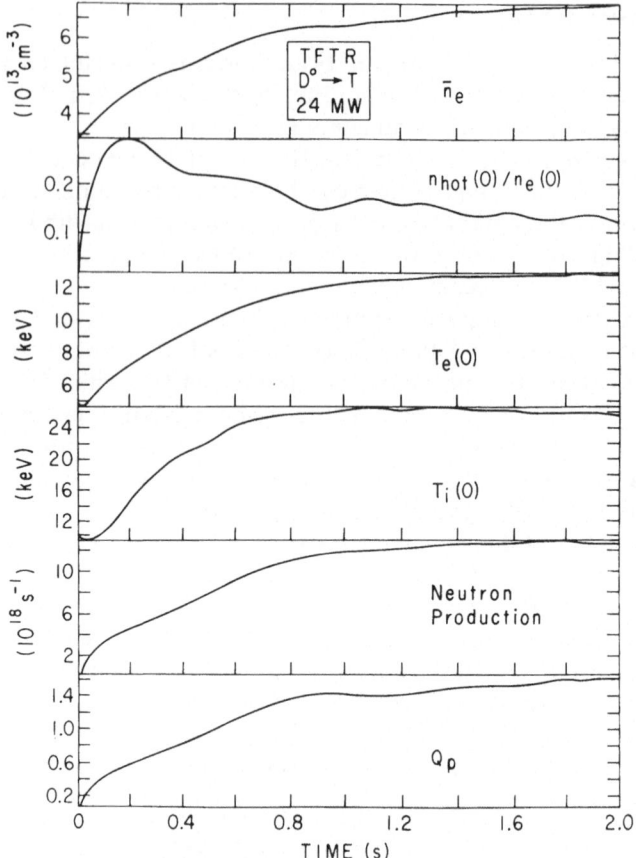

Figure 4.13. Time-dependence of plasma parameters for a $D^0 \rightarrow T$ case (TFTR).

4.2.2.2. *Time-dependence of plasma parameters*

Figure 4.13 shows the time-dependences of several plasma parameters for a case ($D^0 \rightarrow T$) with $P_{beam} = 24$ MW, $D_{0a} = 500$ cm^2 s^{-1}, $D_{1a} = 5000$ cm^2 s^{-1}, $D_{2a} = 5 \times 10^{16}$ cm^{-1} s, $D_{3a} = 0$, $D_h = 0$, $K_{e0} = 5 \times 10^{17}$, $K_{e1} = 0$ and $R_c = 0.75$. The bulk plasma diffusion coefficient is in the middle of the range considered likely to be encountered in practice (see Section 4.2.2.5), and an R_c of this magnitude can be obtained with a large-area gettering system, such as is planned for the TFTR.

The time for the neutron production rate, F_n, to reach a steady value is determined by; (i) the slowing-down time τ_s of the fast ions, which is the time scale for $f_h(v)$ to be formed; (ii) heating of the bulk ions and electrons; and (iii) increase in the plasma density resulting from beam fueling and gas puffing

(1500 A-equivalent in this case). As is evident from Fig. 4.13, $\bar{n}_e$, $T_e(0)$, and F_n (and thus Q_p) reach 90% of their maximum values in about 1 s, and these quantities increase rather slowly after that time. $T_i(0)$ reaches its final value in a somewhat shorter time. Hence a pulse length of only 0.5 s, as originally prescribed for the TFTR (Spano, 1975), would seem to be inadequate for realizing the potential of this machine (assuming the validity of the transport coefficients used here), thereby adding further support for upgrading the pulse length of the injectors. However, it has also been found that the quasisteady values are reached somewhat faster when R_c is increased above 0.75, or when the hot ions are given appreciable radial difusion.

By specifying $T_e(0) = 5$ keV and $T_i(0) = 10$ keV at $t = 0$, the time to reach nearly steady state is reduced by approximately 0.15 s. Most of the results reported in subsequent sections use these starting conditions.

The fraction of hot ion density, n_{hot}/n_e, builds up to a large value at $r = 0$ in about one τ_s, but this quantity decreases to about 0.12 as the bulk plasma density increases. While it is not evident in Fig. 4.13, the total number of thermal deuterons (fueled by the beams and by recycling) becomes equal to the number of thermal tritons (fueled by gas puffing and by recycling) after about 1 s, and n_{Di} considerably exceeds n_{Ti} in the central plasma region.

Table 4.5 gives the principal plasma and fusion parameters for this reference case, at $t = 2$ s. This case has also been followed for 5 s, and is discussed in Section 4.2.2.4. It is shown there that $D^0 \rightarrow T$ operation may be unattractive

Table 4.5. Illustrative parameters at $t = 2$ s for $D^0 \rightarrow T$ operation of TFTR.

Beam energy (D^0)	120 keV
Beam power	24 MW (total)
Tritium gas injection	1500 A-equivalent
Recycling coefficient	0.75
Bulk ion diffusion coefficient (cm^2 s^{-1})	$500 + 5000(r/a)^3 + 5 \times 10^{16}/n_e$
Hot ion diffusion coefficient	$D_h = 0$
$\langle Z_{\text{eff}} \rangle$	1.41 (iron)
$n_e(0)$	1.31×10^{14} cm^{-3}
$n_{\text{hot}}(0)$	1.47×10^{13} cm^{-3}
$\langle n_{\text{hot}} \rangle/\langle n_e \rangle$	0.081
$T_e(0)$	13.0 keV
$T_i(0)$	25.9 keV
$n_{\text{Di}}(0)$	7.79×10^{13} cm^{-3}
$n_{\text{Ti}}(0)$	3.65×10^{13} cm^{-3}
$n_e(0)\tau_E$	2.6×10^{13} cm^{-3} s
$\langle \beta \rangle$	0.019
Fusion Production	
Thermonuclear	9.5×10^{18} n/s
Beam–target	3.9×10^{18} n/s
Total	1.3×10^{19} n/s
Q_p	1.57

for pulse lengths exceeding a few seconds, since both the beam–target and thermonuclear reaction rates decrease as the thermal plasma becomes dominated by deuterium. On the other hand, if pellet injection is successful in directly fueling the central plasma region, then the $D^0 \rightarrow T$ mode might be suitable for very long operating periods.

4.2.2.3. Dependence on beam power

Two modes of operation are compared in the present study. In mode I ($D^0 \rightarrow T$), 120-keV D^0 beams are injected perpendicularly, while the tritium bulk plasma is fueled by recycling ($R_c = 0.75$) and gas puffing. The tritium particle injection rate is 1500 A-equivalent, which corresponds to about 450 Curies s^{-1}, or approximately the maximum specified for TFTR operation (Spano, 1975). In mode II, 100-keV D^0 and 150-keV T^0 beams are injected tangentially with equal currents, but in opposite directions; $R_c = 0.20$ and there is no gas puffing. Thus, mode II can be identified as a CIT-type of operation (Jassby and Towner, 1976; Kulsrud and Jassby, 1976; Jassby et al. 1977).

Previous studies have shown that injection of both D^0 and T^0 beams is especially advantageous at low plasma density, where tangentially injected beams can penetrate easily. Operation with D^0 beams only is more suitable for higher plasma densities, where injection at a more oblique angle is needed for adequate penetration; hence our choice of injection directions for modes I and II.

For all cases reported in this section, the bulk plasma diffusion coefficient is defined by $D_{0a} = D_{3a} = 0$, $D_{1a} = 5000$ cm^2 s^{-1} and $D_{2a} = 5 \times 10^{16}$ cm^{-1} s^{-1}. The energetic ions are assumed to decelerate without displacement from their drift surfaces of birth. Figures 4.14 and 4.15 show the calculated plasma parameters as a function of P_{beam}, after 2 s of injection. In the CIT mode, Q_p rises to values above 0.5 with relatively small P_{beam}, but never becomes much higher than unity. In mode I, Q_p rises linearly with P_{beam} to values near 3 at the highest power investigated. (A recycling coefficient as low as 0.2 is actually not feasible in the TFTR where only a fraction of the wall can be covered by a getter system. However, the low R_c results are applicable to TFTR-sized machines equipped with a magnetic divertor.)

4.2.2.3.1. *Small beam power*. At smaller beam powers, the fusion production in mode I is due overwhelmingly to beam–target reactions, R_{12}. This reaction rate is limited by the relatively low T_e, which reduces the fast ion slowing-down time, and the rate is also degraded by deuterium build-up in the target plasma. In the CIT mode, T_e is somewhat higher because of the much larger ratio of n_{hot}/n_e, and T_i is very much higher because the thermal ions originate mainly from fast ion fueling. The higher T_e and T_i result in a large R_{12}, even though the target composition is much less favorable than that in mode I. Further-

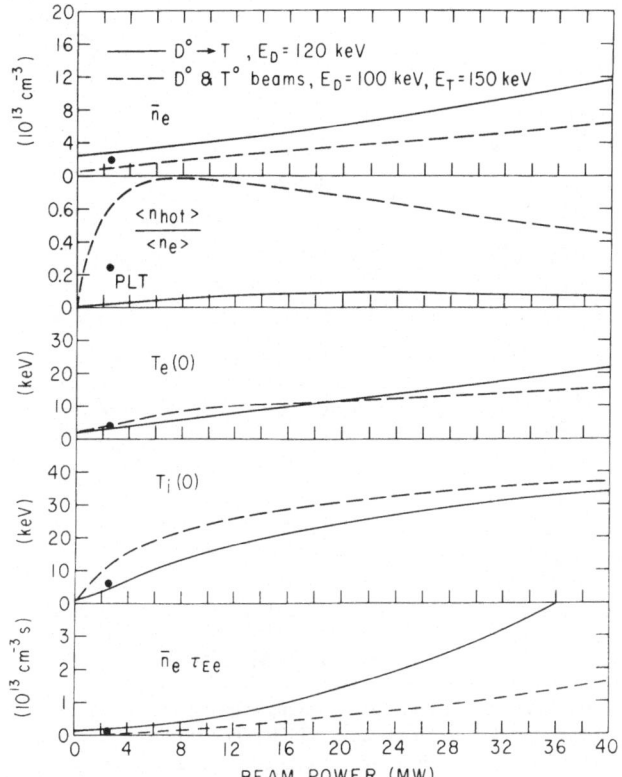

Figure 4.14. Dependence of TFTR plasma parameters on injected beam power for $D^0 \rightarrow T$ with gas puffing, and for injection of D^0 and T^0 beams. Time $= 2.0$ s after injection.

more, the enerergetic ion reaction rate (R_{22}) is substantial in mode II. In both cases, R_{11} is relatively unimportant at low P_{beam}.

4.2.2.3.2. *Large beam power.* At larger beam powers, the reaction rate in mode I increases rapidly because of the larger thermal ion density at increasingly higher T_i. Note that T_e does not decrease with increasing n_e, because of our use of the model $K_e \propto n_e^{-1}$. As R_{11} increases, the fusion alpha production further raises T_e and T_i. Thus with increasing P_{beam}, mode I becomes the BDTN mode (Jassby, 1977). In the CIT mode with small R_c, the total bulk ion energy remains small with increasing P_{beam}, so that R_{12} continues to play a major role.

4.2.2.3.3. *Hot ion pressure.* In mode I, the fraction of hot ion density is always less than 0.1, while in the CIT mode, where essentially all fueling is performed by the beams, this parameter can be as large as 0.8. In mode I, $T_i/T_e \sim 2$, while

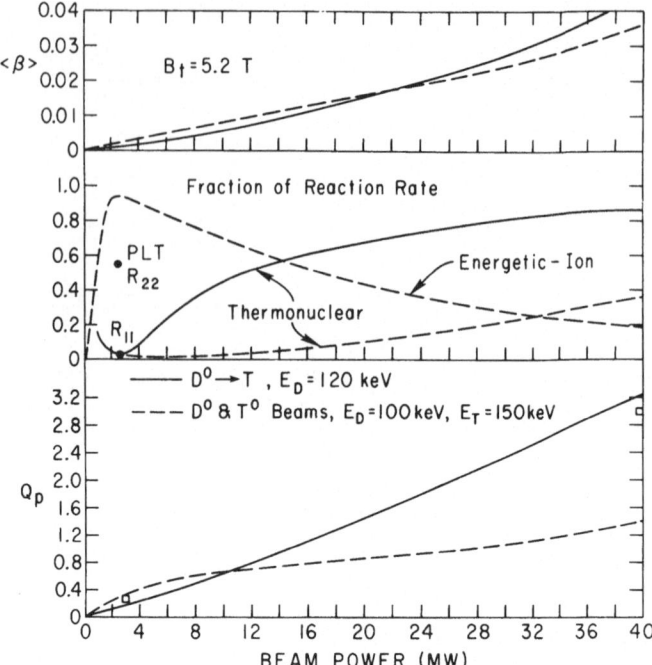

Figure 4.15. Fusion performance and "beta" versus injected beam power for $D^0 \to T$ with gas puffing, and for injection of D^0 and T^0 beams (TFTR). The squares in the graph of Q_p are results for D^0 and T^0 beam injection with a large recycling rate and gas puffing. Time = 2.0 s after injection.

in mode II, $T_i/T_e \sim 3$, the larger ratio being due to the smaller influx of cold ions, which tends to depress T_i. The value of $\langle \beta \rangle$ in mode II is larger than in mode I for P_{beam} up to 15 MW, and a much larger fraction of the pressure in the CIT mode is due to the energetic ions. The largest value of $\langle \beta \rangle$ that can be supported in TFTR (because of structural limitations) is $\langle \beta \rangle \approx 0.04$ (at $B_t = 5.2\ T$), or somewhat smaller than the $\langle \beta \rangle$ that can be attained theoretically with $P_{\text{beam}} \gtrsim 40$ MW.

4.2.2.3.4. *Confinement.* The values of $\bar{n}_e \tau_E$ are much smaller in the CIT mode, in agreement with previous studies (Jassby and Towner, 1976; Kulsrud and Jassby, 1976; and Jassby et al., 1977). While the higher values of $\bar{n}_e \tau_E$ in mode I are in fact consistent with empirical heat transport coefficients, it is reassuring that $Q_p \sim 1$ could still be attained in mode II in the event that unwelcome phenomena such as enhanced radiation or MHD turbulence limit the attainable $\bar{n}_e \tau_E$ to less than 10^{13} cm^{-3} s.

4.2.2.3.5. *Recycling.* If the recycling rate is allowed to increase in mode II operation and if gas puffing is utilized, then Q_p also increases rapidly with

P_{beam}, and becomes close to the values for mode I, as indicated in Fig. 4.15. Thus, at large n_e the reduction in R_{22} as n_{hot} is reduced is compensated by the increased R_{11}. The $\bar{n}_e \tau_E$ values for mode II also increase.

Thus CIT-type operation is advantageous for reaching interesting power multiplications at relatively small P_b and $\bar{n}\tau_E$, but the BDTN mode gives superior performance at very large P_b and $\bar{n}\tau_E$. On the other hand, equally large Q_p can be obtained with both D^0 and T^0 beam injection when R_c is increased and gas puffing is employed, as indicated by the two points in Fig. 4.15. It should be emphasized that the favorable dependences with increasing plasma density observed here arise from the $K_e \propto 1/n_e$ relation, which is assumed to hold under all conditions of density and temperature.

In mode II, the dominant neutron source for $P_{\text{beam}} \leq 20$ MW are the energetic ion reactions, as a result of the large values of n_{hot}/n_e ($\gtrsim 1/2$) that are characteristic of $R_c \ll 1$. These values are higher than have been achieved in PLT thus far ($n_{\text{hot}}/n_e \leq 0.3$), where R_c is somewhat higher and there is significant gas influx from the beam lines. Nevertheless, in PLT plasmas the ratio R_{22}/R_T is as large as 0.5 to 0.6 (Eubank et al., 1979; Strachan, et al., 1981), and R_c is estimated to be about 0.75. The PLT beam injection experiments ($D^0 \rightarrow D^+$) with a maximum P_{beam} of 2.4 MW were found to give the highest Q_p when n_e was small and recycling was minimized (Strachan et al., 1981). As shown by the solid dots in Fig. 4.14 and 4.15, the PLT plasma characteristics at 2.4 MW beam power fall midway between the calculated values for mode I and mode II operation in the TFTR. The fact that the measured PLT parameters actually fall in the ranges expected from the code results indicates that reasonable agreement between the code predictions and experimental results on the TFTR should be realizable.

4.2.2.4. Radial profiles and long pulses

Figure 4.16 shows the radial profiles of plasma densities and temperatures at $t = 2$ s for a mode I case ($R_c = 0.75$) and a mode II case ($R_c = 0.2$), both with 24 MW injection. The relatively flat T_i profile for mode II is characteristic of a plasma with $R_c \ll 1$, such as can be obtained by intensive gettering (in the TFTR) or by an "unload" divertor. The total plasma density, n_e, is larger in mode I, and has a slightly broader profile; these effects result both from gas puffing and from the higher value of R_c in mode I operation.

Important differences are evident in the radial profiles of warm ion density, n_{Di} and n_{Ti}. In mode II, $n_{\text{Ti}}(r) \approx n_{\text{Di}}(r)$, which is expected because both plasma species are fueled equally intensively by neutral beam injection. For mode I, on the other hand, n_{Di} exceeds n_{Ti} by a factor of the order of two in the high-temperature region of the plasma, an effect that is due to beam fueling of the warm deuterons. This phenomenon drastically reduces the relative importance of beam–target reactions (26% in mode I, compared with 50% in mode II, at $t = 2$ s), and also reduces the thermonuclear reaction rate.

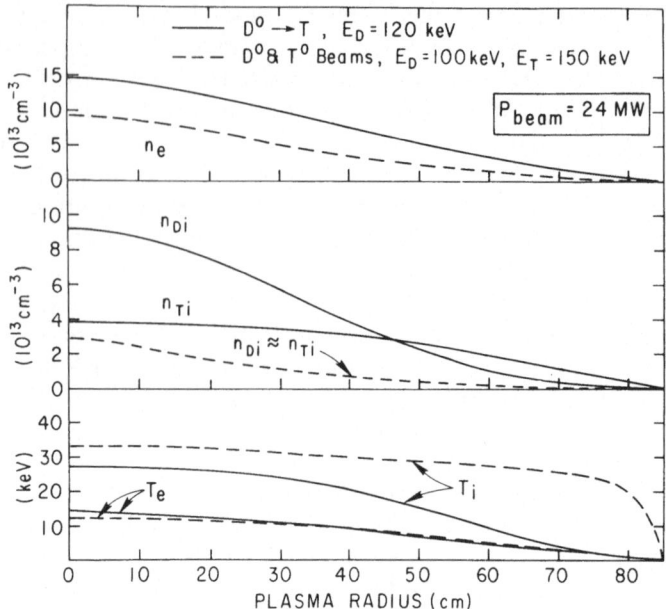

Figure 4.16. Radial profiles of plasma densities and temperatures for $D^0 \to T$ with gas puffing, and for injection of D^0 and T^0 beams (TFTR). Time = 2.0 s after injection.

Figure 4.17 shows the time-dependence of Q_p for mode I and for a modified mode II operation with pulse lengths up to 5 s. Each case has $R_c = 0.75$ and 1500 A-equivalent gas injection; in mode II, equal gas currents of D and T (as well as beams) are injected. (The mode I parameters are the same as for Fig. 4.16, except for a slightly different diffusion coefficient.) In mode I, with D^0 injection only, Q_p reaches its maximum value at $t = 2.7$ s, then drops continuously, and by $t = 5$ s it has decreased below the Q_p obtained in mode II. This effect is due to increasing depletion of tritium in the central region, where the fusion reactivity is highest. Thus for pulse lengths of practical reactor interest ($\gtrsim 30$ s), it is necessary to use both D^0 and T^0 beam injection to maintain $Q_p \approx$ constant, or else to develop an effective means of pellet injection to supply tritium to the central plasma region (Jassby and Towner, 1976).

4.2.2.5. *Effect of particle diffusion*

4.2.2.5.1. *Variation of the bulk plasma diffusion coefficient.* The thermal conductivities of electrons and ions in tokamak plasmas are presently described by empirical scalings (or by approximately neoclassical values for the ions). These scalings will doubtless be modified in the future, but at least there is some reasonable justification for them on the basis of the behavior of present-day tokamak plasmas. On the other hand, the magnitude and parametric

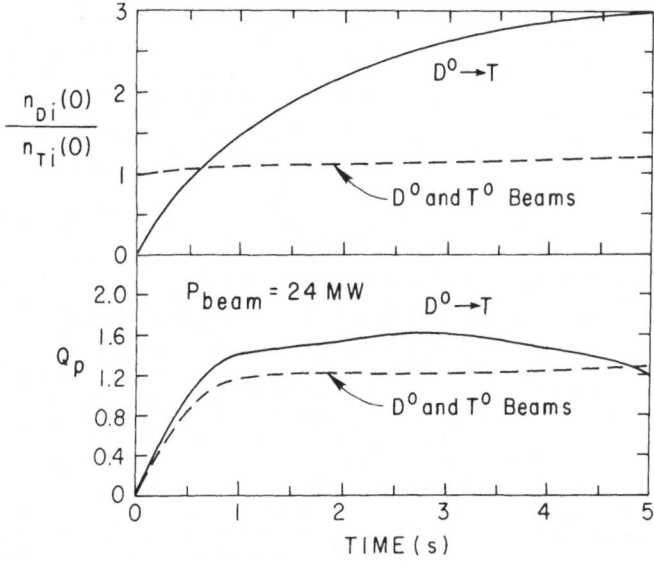

Figure 4.17. Time-dependence of the thermal ion composition and fusion performance for $D^0 \to T$ and for injection of D^0 and T^0 beams (TFTR). Each case has $R_c = 0.75$ and gas injection.

dependences of the experimental particle diffusion coefficient is far less soundly known, with empirical scalings being modified rather frequently. In order to investigate the effect of various diffusion coefficients on the performance of the TFTR plasma, $D_{1a} = 5000$ cm^2 s^{-1} and $D_{2a} = 5 \times 10^{16}$ cm^{-1} s^{-1} have been used as in previous sections, but with D_{0a} varied over a wide range. The results are shown in Fig. 4.18 for $P_{\text{beam}} = 24$ MW and 4 MW with deuterium beam injection only (i.e., $D^0 \to T$).

With increasing diffusion rate, $\bar{n}_e$ decreases as expected. $T_i(0)$ remains high because of the important effect of beam fueling, but $T_e(0)$ drops appreciably, because of less effective coupling to the thermal ions.

Evidently Q_p drops markedly with increasing diffusion rate. At large values of D_0 there is little increase in Q_p with beam power, which indicates the difficulty in producing a large thermal-ion energy-density when convection is extremely fast. The performance could be improved by increasing the tritium gas injection rate and hence the plasma density, but at the penalty of increased tritium inventory. Thus in order to make a confident prediction of the attainable values of Q_p in the TFTR (or in other beam-driven tokamaks), it appears essential to obtain some definitive information from present experiments on the magnitude of the particle diffusion coefficient at high temperature.

4.2.2.5.2. *Radial diffusion of hot ions.* The results of present day beam-injected tokamak experiments have shown that the thermalization rate of injected fast

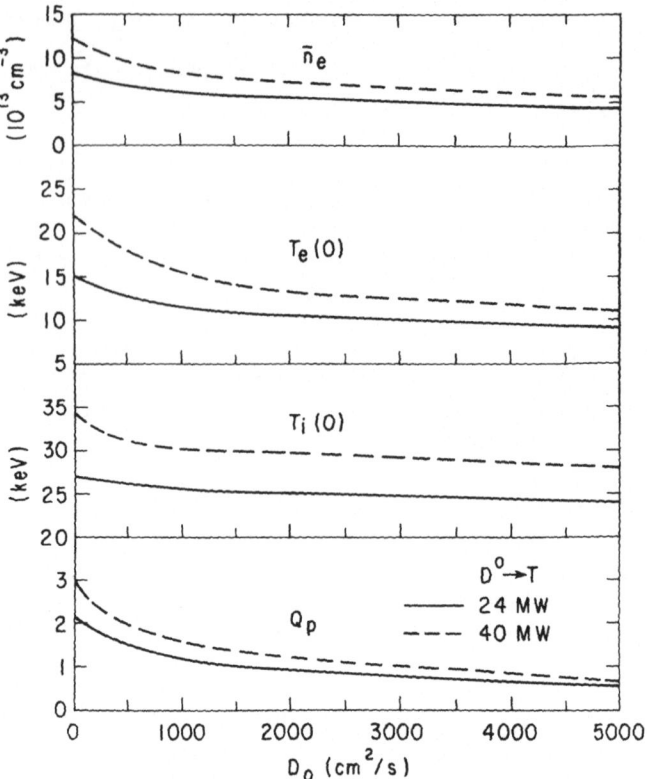

Figure 4.18. Dependence of plasma parameters on thermal ion diffusion coefficient, $D = D_{0a} +$ $5000\,(r/a)^2 + 5 \times 10^{16}\,n_e^{-1}$(TFTR).

ions is classical, within the uncertainties of the experimental data (Jassby, 1977; Eubank *et al.* 1979; Strachan *et al.* 1981). No information is available on the radial diffusion of the fast ions; one might expect that the fast ions do experience a fraction of the anomalous spatial diffusion characterizing the bulk ions. (Note that while thermalization of the bulk ions and electrons is classical—i.e., the plasma resistivity is known to be classical—anomalous spatial diffusion of the bulk ions still occurs.) Fast ions tend to spatially average the turbulent fields usually present in the tokamak plasma, so that one would expect them to be subject to less radial diffusion than are the thermal ions, if plasma turbulence in fact contributes significantly to radial diffusion. However, in high-temperature TFTR plasmas, the hot ions have a velocity $\sim 3 \times 10^8$ cm s^{-1}, which is typically only about 2.5 times a typical warm ion velocity in the central plasma region where the bulk of the fusion reactions occur. Thus it is reasonable to expect that the fast ions are subject to a significant fraction of the turbulence-induced spatial diffusion experienced by

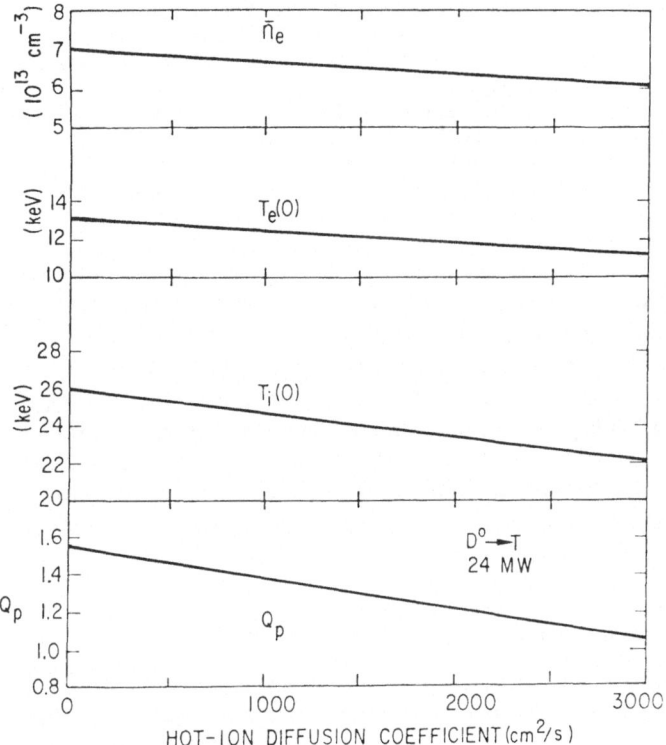

Figure 4.19. Dependence of plasma parameters on hot ion diffusion coefficient (TFTR).

the thermal ions. On the other hand, if neoclassical diffusion is the dominant radial transport mechanism, then one would expect the hot ion diffusion rate to be several times smaller than that of the bulk ions—but perhaps still significant.

Previous theretical analyses of beam-injected tokamaks usually have assumed that the fast ions remain on their drift surfaces of birth while slowing down. The present analysis investigates the effect on fusion performance of a finite radial diffusion rate of the fast ions.

A number of calculations have been performed for the conditions shown in Table 4.5, but with a constant energetic ion diffusion coefficient $D_h > 0$. The results are displayed in Fig. 4.19. As D_h increases from 0 to 3000 cm^2 s^{-1}, Q_p drops from 1.57 to 1.04. This reduction in Q_p results in part from a reduction in the thermonuclear reaction rate; both $\bar{n}_e$ and $T_i(0)$ are reduced, since the hot ions give up a smaller fraction of their energy to the central plasma region. The remainder of the reduction in Q_p is due to the smaller beam–target reaction rate, because the hot ions find themselves in a region where the slowing-down time is smaller.

Table 4.6. Effect of enhanced ion heat
conductivity in TFTR.

K_i/K_{nc}	1	3	10
$\bar{n}_e$, 10^{13} cm^{-3}	7.0	6.9	6.8
$T_e(0)$, keV	13	12	11
$T_i(0)$, keV	26	24	19
Q_p	1.57	1.43	1.15

These results suggest the importance of determining from present beam injection experiments at low density whether hot ion diffusion is in fact significant. From the known behavior of energetic ions in present day beam-injected tokamaks such as PLT (Strachan *et al.*, 1981), it seems unlikely that D_h is as large as the bulk plasma diffusion coefficient, D_a. If D_h is taken as lying in the same range as D_a in the center half of the plasma (typically 500 to 1500 cm^2 s^{-1}), then Fig. 4.19 shows that Q_p is reduced by up to 20% for our reference case. The effect of a nonzero D_h will be larger in low-density plasmas where the slowing-down time is longer.

4.2.2.6. *Variation of the ion thermal conductivity*

Analyses of the performance of beam-heated plasmas in PLT (Eubank *et al.*, 1979), and ISX-B (Swain *et al.*, 1979) have given values of the ion thermal conductivity, K_i, which are a factor of 1 to 3 times the neoclassical value, K_{nc}. (The uncertainties in the experimental determinations are due to the competing-ion energy-loss processes of charge exchange and Coulomb interaction with the plasma electrons.) In order to determine the effect on the TFTR fusion performance of higher K_i, calculations have been performed with the basic parameters of Table 4.5, but with K_i varying up to ten times K_{nc}.

The results are given in Table 4.6. Only for $K_i > 3K_{nc}$ does $T_i(0)$ drop appreciably, thereby producing a marked reduction in the thermonuclear reaction rate and in Q_p. There appears to be little effect on Q_p for the range of K_i/K_{nc} that is measured in beam-injected tokamak experiments.

4.2.2.7. *Variation of the plasma recycling coefficient*

A number of mode I cases with D^0 beam injection alone at P_{beam} = 24 MW and 40 MW were repeated with R_c increased form 0.75 to 0.90. When $D_0 = 0$, Q_p is increased by 25–35%. When $D_0 = 5000$ cm^2/s^{-1}, Q_p is doubled from its value at $R_c = 0.75$ (which is very small, as shown in Fig. 4.18).

The effect of increasing R_c is to increase n_e. In the present transport model, where $K_e \propto 1/n_e$, a larger n_e does not result in a lower T_e. While K_i does increase with n_e, ion heat conduction is usually small compared with convection, so that T_i does not decrease either. Thus the thermonuclear reaction rate increases with R_c. Furthermore, the increased fusion alpha-particle production actually tends to raise T_e and T_i, thus leading to an even higher Q_p. At very large D_0, beam injection becomes the dominant fueling process, so that in mode I operation the ratio n_{Ti}/n_{Di} tends to decrease. As R_c is increased, beam fueling becomes less overriding and the species ratio becomes more favorable both for thermonuclear and beam–target reactions.

The beneficial effect of increasing R_c is reduced at smaller beam powers and moderate D_0. The effect is expected to disappear or become unfavorable when n_e becomes so large as to hinder beam penetration. It is planned to exercise some control over R_c in the TFTR by utilizing getter collection systems to trap the outgoing plasma, a procedure that has proven to be effective in present-day experiments (Eubank *et al.*, 1979; Strachan *et al.*, 1981).

4.2.3. Divertor injection tokamak experiment (DITE)

Simulations of the DITE tokamak (Paul *et al.*, 1977) are now presented. Comparisons between code and experiment are carried out for six well-documented cases, with good overall agreement being obtained. Particular attention is paid to modeling the effect of beam impurity charge exchange on the radiative power loss.

4.2.3.1. *Plasma parameters*

Typical DITE parameters are displayed in Table 4.7. In each of the six cases of interest (see Table 4.8), approximately 1 MW of neutral beams are co-injected into an ohmically heated plasma, with 77% of the beam power at full energy, 18% at one-half energy and 5% at one-third energy. There is no radial diffusion of the fast ions. The recycling coefficient is $R_c = 0.9$, and the gas puffing rate is determined dynamically to match the experimentally measured line density. At the limiter radius, n_e, T_e, and T_i have characteristically small values. The initial conditions correspond to those of a plasma which has already undergone some ohmic heating.

Table 4.7. DITE parameters.

Major radius	1.17 m
Minor radius	0.26 m
B-Toroidal	13–22 T
Plasma current	200 kA
Neutral beam energy	30 keV
Neutral beam power	1 MW (co)
	77% at full energy
	18% at one-half energy
	5% at one-third energy
Recycling coefficient R_c	0.9

Table 4.8. DITE Discharges.[a]

	A	B	C	D	E	F
$B_\phi(kG)$	20	20	22	13.5	13.5	20
$I_G(kA)$	150	160	100	160	110	150
$V_l(v)$	3.8	1.6	1.8	1.6	1.3	1.2
Z_{eff}	3.9	1.4	1.1	2.0	1.0	1.6
$T_e(eV)$	500	650	650	700	450	800
$T'_e(eV)$	500	1200	900	800	800	1050
$T_i(eV)$	300	210	220	200	190	250
$T'_i(eV)$	760	400	650	600	410	590
$\bar{n}_e$	1.2	3.4	1.6	1.0	3.3	2.8
$\bar{n}'_e(\times 10^{13})$	1.6	4.6	2.2	6.0	6.3	3.6
$P_{NI}(MW)$	0.9	0.8	1.1	0.62	0.68	1.0
Gas	H	D	D	D	D	D

[a] Table with main experimental parameters for discharges A–F. The two entries for some parameters (e.g., T_e, T'_e) are the values before and during neutral injection. The temperatures all correspond to $r = 0$.

4.2.3.2. Impurities and radiation

Before neutral injection, the radiated power from the plasma is calculated from the impurity concentration n_z and a coronal equilibrium model (Post *et al.*, 1977): the impurities are not allowed to diffuse. At the onset of neutral injection it is experimentally observed that the radiated power exhibits a rapid and often large increase which is at least partially due to the enhanced radiated power associated with charge-exchange recombination of the injected energetic ions with the highly stripped plasma impurity ions. In the present study the radiated power during injection has been modeled in one of two ways.

(I) It is assumed that during injection the radiated power is enhanced by a factor γ over that given by the coronal equilibrium model. γ is chosen to give agreement with experiment. Although this model is rather arbitrary, it also represents a way of modeling impurity influx into the plasma. Experimentally it is apparent that the impurity concentration of the plasma sometimes rises during injection. Ideally, the impurity influx should be measured and provide an input to the theoretical model. However, these measurements do not exist for the DITE discharges under study.

(II) The charge-exchange recombination enhancement proposed by Hulse *et al.* (1980) is used.

In order to calculate the radiated power terms it is necessary to know the concentration of the various impurities (n_z/n_e) in the plasma. Although these can be obtained in principle from experimental spectroscopic data and then input to the code, for the discharges under consideration this information is generally not available. However, as experimental values of Z_{eff} and the radiated power $P_{rad}(r)$ are available, n_z/n_e is determined from this data, without injection, by assuming that only one species of heavy impurity (Fe) and one light impurity (O) is present. The Fe concentration is determined from the central value of the radiated power and the O concentration is then adjusted to given the correct Z_{eff}. This gives approximately the correct radiated power profile, although a further adjustment is made to n_{Fe} and n_O if there is a large discrepancy. The volume-averaged radiated power is within a factor of 2 of the experimental value in the worst case. It is then assumed that these values of (n_z/n_e) are constant throughout the plasma and at all later times. It is realized that this procedure is not entirely satisfactory, but, in view of the experimental uncertainties, it is adequate for the present purposes.

4.2.3.3. Transport model

The transport coefficients are determined empirically and their initial values are determined as described below.

4.2.3.3.1. *Electron thermal conductivity.* If it is assumed that the electron energy containment time (τ_e) is determined entirely by electron conduction losses, then the following simple power balance determines K_e from the energy containment time and plasma energy (W_e)

$$\frac{W_e}{\tau_e} = K_e n_e \frac{\partial T_e}{\partial r} 2\pi R 2\pi a, \tag{4.2.2}$$

where τ_e is determined from a scaling law. If Alcator scaling (Agpar *et al.*, 1977) $(\tau_e = 4.0 \times 10^{-19} \bar{n}_e a^2$ s) is used then K_e is determined as $K_e = 7.5 \times 10^{17} n_e^{-1}$ cm^2 s^{-1} with n_e in units of cm^{-3}. The modified Alcator scaling (Pfeiffer and Waltz, 1979), which includes a dependence on the plasma major radius, gives $K_e = 3.4 \times 10^{17} n_e^{-1}$ for DITE. Our initial choice is therefore $K_{e0} = 5 \times 10^{17}$ and $K_{e1} = 0$ (see (4.1.19)).

4.2.3.3.2. *Ion thermal conductivity.* Although several studies have shown that the experimental data in some machines is consistent with neoclassical ion heat conduction, a previous analysis (Gill, 1979) of a high-power injection heated discharge in DITE showed that the ion thermal conductivity exceeded the neoclassical value K'_{nc} by a factor of about 5, giving $K_i = 5 \times K'_{nc}$.

The value used for K'_{nc} is similar to that given in (4.1.18) and may be expressed as follows:

$$K'_{nc} = K_i \text{(plateau)} \frac{0.36 v_i^*}{1 + 0.36 v_i^*}, \tag{4.2.3}$$

with

$$K_i \text{(plateau)} = \frac{3}{2}\sqrt{\frac{\pi}{2}} \frac{v_i q \rho_{i\phi}^2}{R}, \tag{4.2.4}$$

Table 4.9. Ion thermal conductivity in DITE.

Discharge	v_i^*	K_i (plateau) (cm s^{-1})	K'_{nc} (cm^2 s^{-1})
A before NI[a]	0.63	8.1×10^3	1.5×10^3
during NI	0.13	3.3×10^4	1.5×10^3
B	3.44	6.4×10^3	3.5×10^3
	1.32	1.6×10^4	5.3×10^3
C	2.62	1.0×10^4	4.9×10^3
	0.48	4.5×10^4	6.7×10^3
D	0.75	8.6×10^3	1.8×10^3
	0.49	4.5×10^4	6.7×10^3
E	4.55	1.1×10^4	6.8×10^3
	1.67	3.8×10^4	1.4×10^4
F	2.11	8.7×10^3	3.8×10^3
	0.48	3.2×10^4	4.7×10^3

[a] NI stands for neutral injection.

and collisionality parameter

$$v_i^* = \frac{R^{3/2}}{r} \frac{qR}{v_i \tau_i}, \tag{4.2.5}$$

where q is the plasma safety factor, v_i is the ion velocity, $\rho_{i\phi}$ is the ion Larmor radius in the toroidal field, and τ_i is the ion collision time. These are defined as

$$v_i = \sqrt{\frac{T_i}{m_i}}, \qquad \rho_{i\phi} = \frac{c\sqrt{2T_i m_i}}{Z_i e B_\phi}, \tag{4.2.6}$$

$$\tau_i = \frac{3m_i^{1/2} T_i^{3/2}}{4\pi^{1/2} n_i Z_i^4 e^4 \ln \Lambda}, \tag{4.2.7}$$

where m_i is the ion mass and $\ln \Lambda$ is the Coulomb logarithm. Both plateau and neoclassical ion thermal conductivities together with v_i^* are evaluated for the DITE discharges at $r/a = 0.5$ and are shown in Table 4.9. Our value of K'_{nc} is between the values of Hinton and Hazeltine (1976) and Hinton and Rosenbluth (1973) for the cases considered here.

4.2.3.3.3. *Diffusion coefficient.* Previous experimental data (Hugill, 1983) suggests that the particle diffusion coefficient can be up to a factor of 10 less than K_e. Alternatively, a simple model can be used to derive D from the neutral density data determined by five different methods. This leads to the formula (see (4.1.13)) $D_{2a} = 1.6 \times 10^{16}$, $D_{0a} = D_{1a} = D_{3a} = 0$ (Gill et al., 1984).

4.2.3.3.4. *Modifications to the above.* The initially chosen set of transport coefficients gives results in reasonable agreement with experiment. However, the calculated electron temperatures, both before and during injection, are too low and K_{e0} is therefore decreased to 2×10^{17} to give better agreement.

The calculated ion temperatures during injection are initially too high and are therefore reduced by increasing the ion conductivity enhancement factor to 10. Although this produces an improved agreement, the central ion temperature is rather insensitive to this factor. Even after this change, the calculated ion temperature for discharge A (see Table 4.8) during injection is still too large and the calculated neutral density too small. In addition, the electron density profiles are too peaked during neutral injection. Because discharge A is the lowest density discharge considered, both of these problems are overcome by using the modified diffusion coefficient below

$$K_{e0} = 2 \times 10^{17}, \tag{4.2.8}$$

$$K_i = 10 \times K'_{nc},$$

$$D_{0a} = \frac{3.9 \times 10^8}{\bar{n}_e^{0.43}},$$

$$D_{2a} = \frac{2.7 \times 10^{22}}{\bar{n}_e^{0.43}},$$

where $\bar{n}_e$ is the line-averaged electron density. The diffusion coefficient at low and high densities is then:

(i) Low $\qquad \bar{n}_e = 2 \times 10^{13}$ cm^{-3}, $\hfill$ (4.2.9)

$$D = \left(740 + \frac{5.1 \times 10^{16}}{n_e}\right) \text{cm}^2 \text{ s}^{-1}.$$

(ii) High $\qquad \bar{n}_e = 6 \times 10^{13}$ cm^{-3}, $\hfill$ (4.2.10)

$$D = \left(461 + \frac{3.2 \times 10^{16}}{n_e}\right) \text{cm}^2 \text{ s}^{-1}.$$

4.2.3.4. Simulation results

In the initial calculations, radiation model I gives the results shown in Table 4.10. It is seen that, with the exception of discharge F before injection, where the calculated average $\langle T_e \rangle$ exceeds experiment by 60%, very good overall agreement is achieved for the temperatures both before and during injection. The total radiated power $\langle P_{rad} \rangle$ is in poorer agreement, but because of the large experimental errors on P_{rad}, there is little to be gained at present in trying to model these quantities more accurately.

The ion temperatures, while in reasonable overall agreement, show two systematic effects. (i) During injection, for the low-density discharges T_i^{calc} exceeds T_i^{exp} with the opposite occurring in high-density discharges. (ii) Before injection the calculated values of T_i are systematically too low. In fact, the ion temperatures before injection are best simulated with an ion conductivity enhancement factor of 4.

Table 4.10. Summary of DITE simulation results.

Discharge		$\langle T_e \rangle^a$ (eV)	$\langle T_e \rangle^b$ (eV)	T_i^a (eV)	T_i' (eV)	$\langle P_{rad} \rangle$ (kW)	$\langle P_{rad}' \rangle$ (kW)	P_Ω^c (kW)
A	Experimental	253	265	300	760	142	392	570
	Theory	276	274	200	980	72	540	
B	Experimental	225	318	210	395	73	298	256
	Theory	188	355	200	480	120	198	
C	Experimental	116	215	220	600	29	310	180
	Theory	144	353	130	810	21	168	
D	Experimental	289	313	200	600	—	279	256
	Theory	306	299	120	440	—	291	
E	Experimental	132	249	180	410	—	—	143
	Theory	119	270	130	380	—	—	
F	Experimental	265	317	250	590	94	251	180
	Theory	160	281	200	740	97	125	
Mean difference %		21	9	40	21	40	53	
Mean ratio		1.11	0.93	1.42	0.93	1.23	1.40	

[a] Values of T_e, T_i, etc. are those before (unprimed) and during (primed) neutral injection.
[b] $\langle T_e \rangle = \int T_e r \, dr / \int r \, dr$. T_i is the value at $r = 0$.
[c] P_Ω is determined before injection.

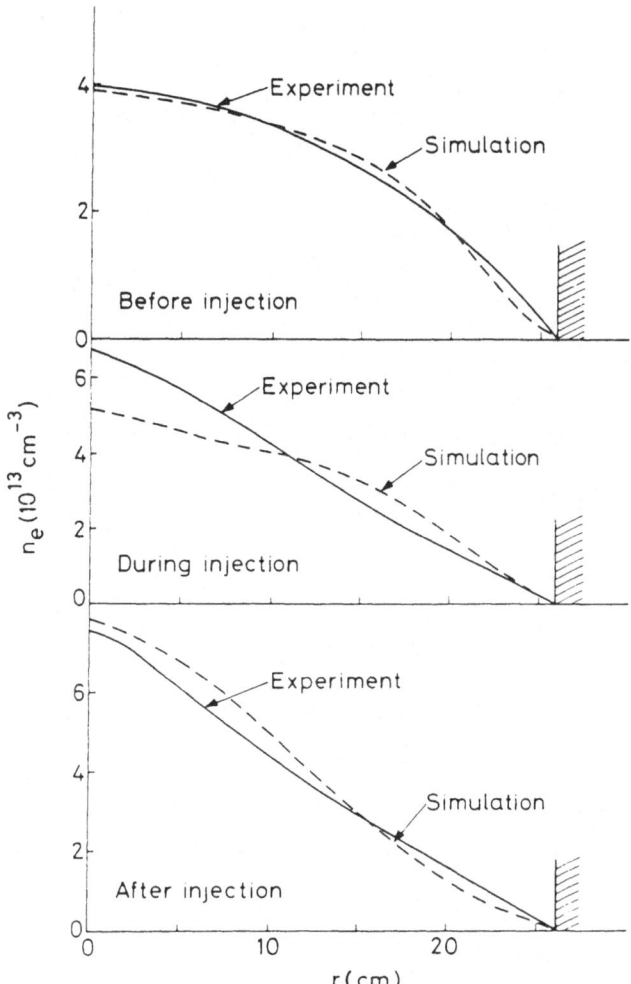

Figure 4.20. Electron density radial profiles before, during, and after neutral injection for DITE discharge F.

Despite the various uncertainties, the values obtained in Table 4.10 represent the best overall agreement which can be achieved without making many very detailed changes in the transport coefficients intended to improve the agreement in specific cases. It is not believed that this would produce an overall improvement in our ability to model other DITE discharges.

Some of the typical profiles are shown in Figs. 4.20–4.23. In general, $n_e(r)$ is in good agreement with calculation for all discharges and it is interesting that for discharge F (Fig. 4.20) the changes in profile before, during, and after injection are well modeled. The shape of the electron temperature profile is also in

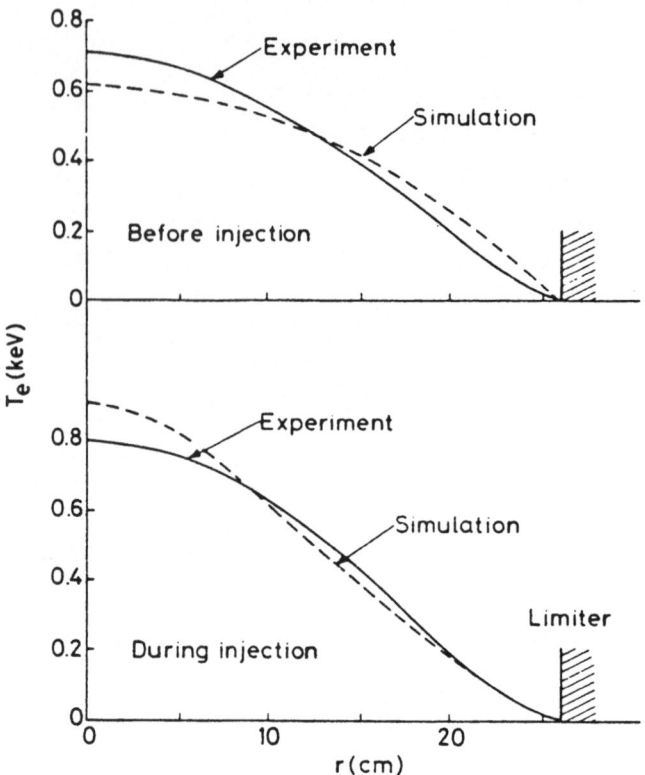

Figure 4.21. Electron temperature profiles before and during injection for DITE discharge D.

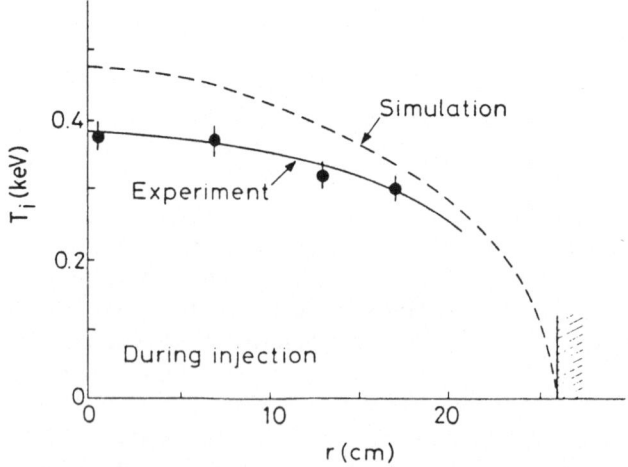

Figure 4.22. Ion temperature profile for DITE discharge B.

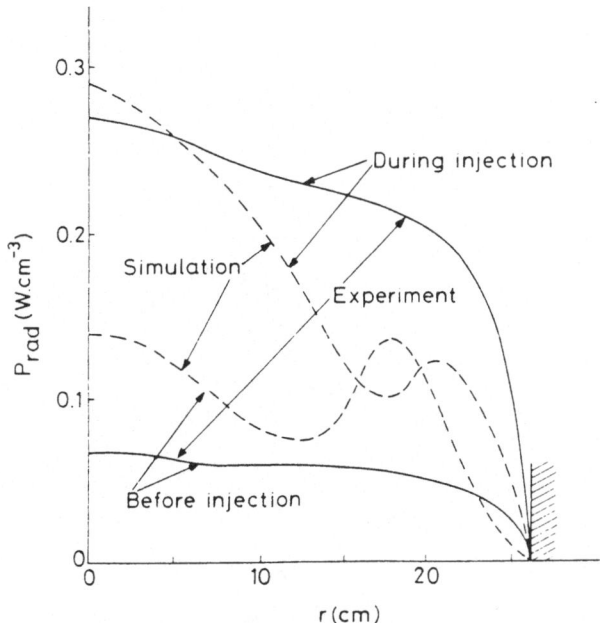

Figure 4.23. Radiated power profiles before and during injection for DITE discharge B.

good agreement (Fig. 4.21). The experimental data on the ion temperature profiles is somewhat inadequate because measurements of T_i are generally available only out to $r = 18$ cm. The profiles are very flat, varying typically as $[1 - (r/a)^3]$. In the example shown (Fig. 4.22) the shape of $T_i(r)$ is well reproduced although the calculated magnitude of T_i is slightly too large.

A typical radial dependence of the radiated power is shown in Fig. 4.23 before and during injection. The central peak in the calculated curve is caused by radiation from Fe whereas the peak at $r \approx 20$ cm is caused by radiation from O. In the experimental curves this structure is eliminated, presumably because of contributions from impurities other than Fe or O.

Finally, in Fig. 4.24, the comparison is made for neutral density. The full line represents the average of the experimental values and the simulated values of n_0 at $r = 0$ and 18 cm are shown. The agreement at $r = 0$ is superficially excellent but it must be remembered that the experimental values probably represent averages over the central volume of the plasma.

The importance of the radiated power and the accuracy of the modeling is checked by comparing simulations with radiation models I and II, all other things being equal. In Table 4.11 the electron temperature and total radiated power are shown for the two models for discharges A to F. Except for case A, the electron temperature does not vary greatly. It is clear that for discharge A the very large radiated power during injection effectively keeps T_e down at a

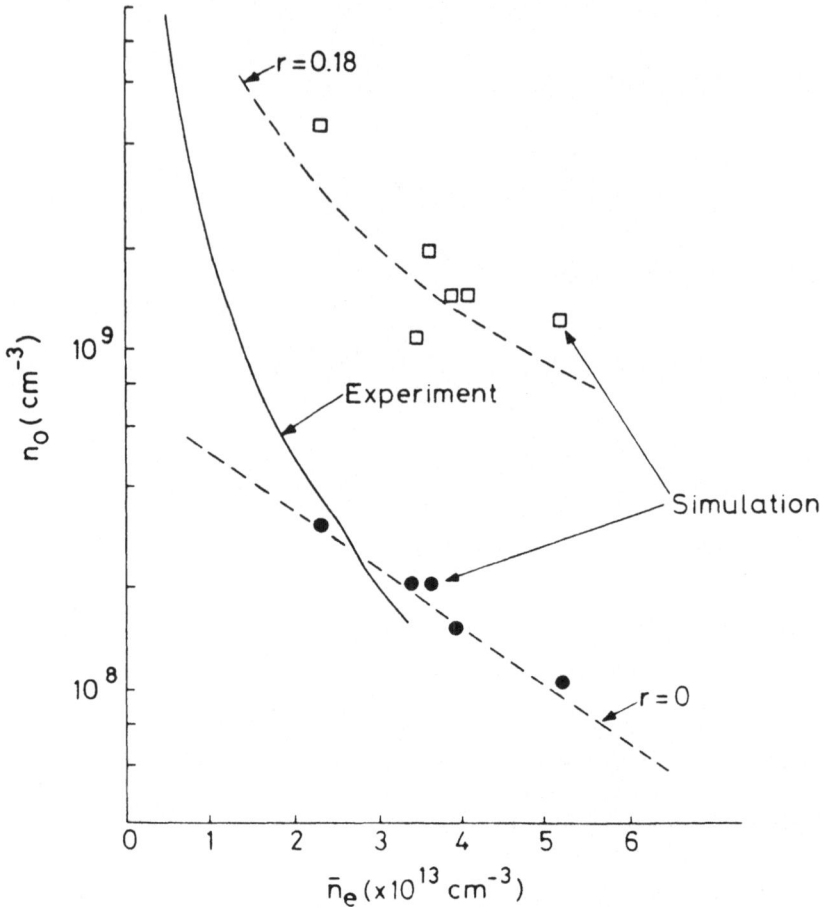

Figure 4.24. Simulated and experimental neutral density for D_2 DITE discharges.

Table 4.11. Radiation models for DITE simulations.[a]

	T_e			Radiated Power		
Model	I	II	Experimental	I	II	Experimental
Discharge						
A	550	1300	500	540	189	392
B	1200	1300	1200	198	50	298
D	1100	1100	800	291	265	279
E	900	900	800	—	—	—
F	1200	1200	1050	125	129	251

[a] Electron temperature and radiated power for the different radiation models during neutral injection.

low level. For discharges D and F the radiated power is adequately simulated by both models, but for discharges A and B the large differences between models I and II imply an influx into the discharge during injection of impurities and this is not modeled in the present calculation. This is confirmed in an approximate way for discharge B by the rising plasma loop voltage during neutral injection at a time when T_e is also rising. Discharge C presents further difficulties because of the apparent very large observed rise of radiation at the onset of neutral injection. This seems to have been caused by experimental uncertainties and therefore a comparison has not been made in this case.

4.2.3.4.1. *Collapse of discharge* D. In discharge D the electron density increases to very nearly the disruptive limit. The behavior of the simulation as this limit is approached is therefore of particular interest. The time-dependence of $\bar{n}_e$ and central values of T_i and T_e are all shown in Fig. 4.25. From the simulation it is clear that a collapse of T_e and T_i starts when the electron density reaches $\sim 4 \times 10^{13}$ cm^{-3} at 140 ms. This is caused by the dominance in the power balance of radiation (see Fig. 4.26) and ion losses which both increase further as the density increases, causing the discharge to be extinguished in the simulation and leading to a disruption in the experiment. It is unfortunate that more detailed measurements of T_e do not exist; the ones shown, although in agreement with the simulation, are unable to confirm the detailed downward trend at $t > 150$ ms. However, the general shape of $T_i(t)$ is in good agreement with the simulation. This behavior was not observed for the rather similar discharge E because of the lower Z_{eff} are reduced role of radiation.

4.2.3.5. Uncertainties

The fact that reasonable overall agreement is obtained between the theoretical model and the experimental data leads to the presumption that the model is basically correct. However, whether all aspects of the model are correct in detail, or whether the diffusion coefficients and conductivities are accurately determined are more complicated questions. One problem with the theoretical model is that some quantities such as the plasma neutral density and the plasma radiated power are known to have toroidal variations which cannot adequately be described by a one-dimensional model.

For the present data set some indication of the probable toroidal variations of n_0 can be obtained by comparing values of n_0 determined by local measurements with those determined by methods which are toroidally averaged. It can be shown (Gill *et al.*, 1982) that this effect is not more important than the other uncertainties in the determination of n_0, and to within perhaps a factor of 2 the toroidal variation in n_0 can therefore be ignored.

It is also known (Hulse *et al.*, 1980) that the injection of a fast neutral beam causes locally enhanced radiative emission due to the charge-exchange recombination of the beam with the highly stripped impurity atoms. Although our radiation model II includes this effect, the one-dimensional nature of our cal-

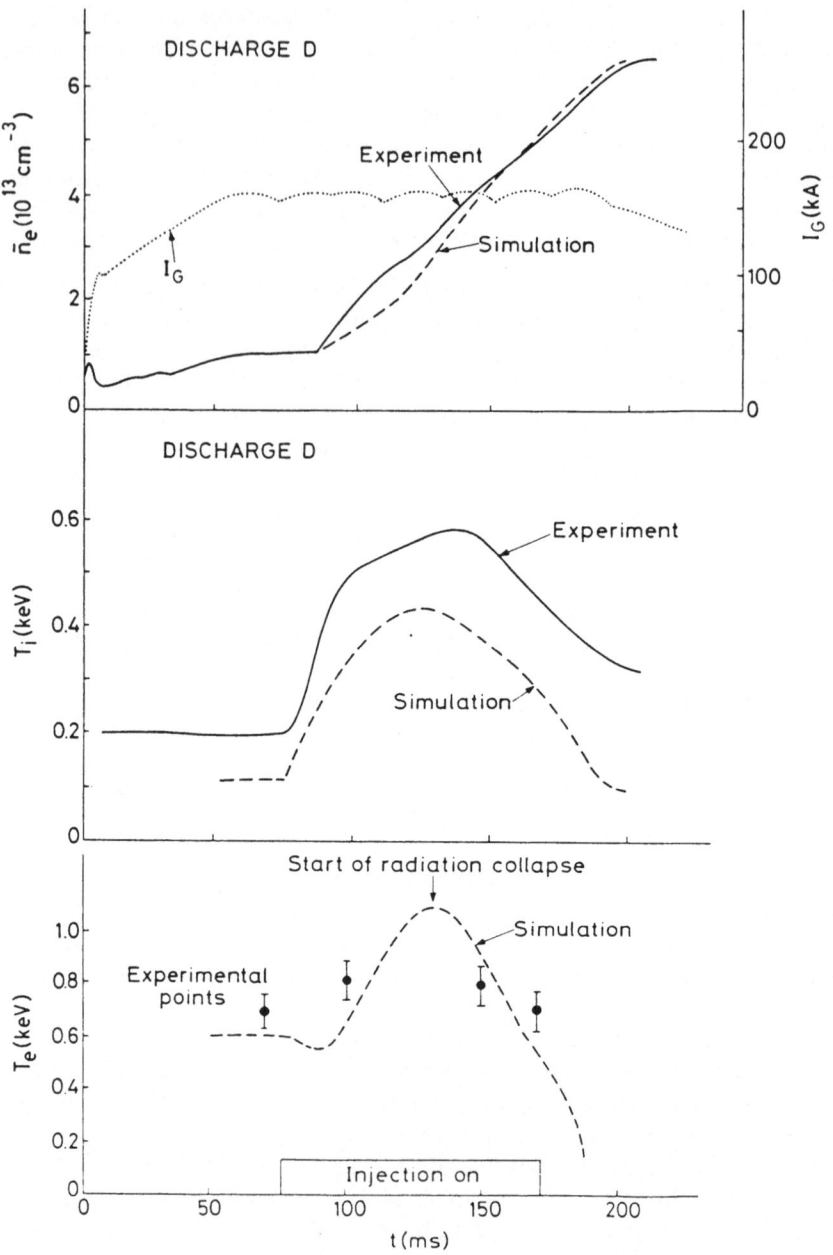

Figure 4.25. Simulation of the radiation collapse of DITE discharge D.

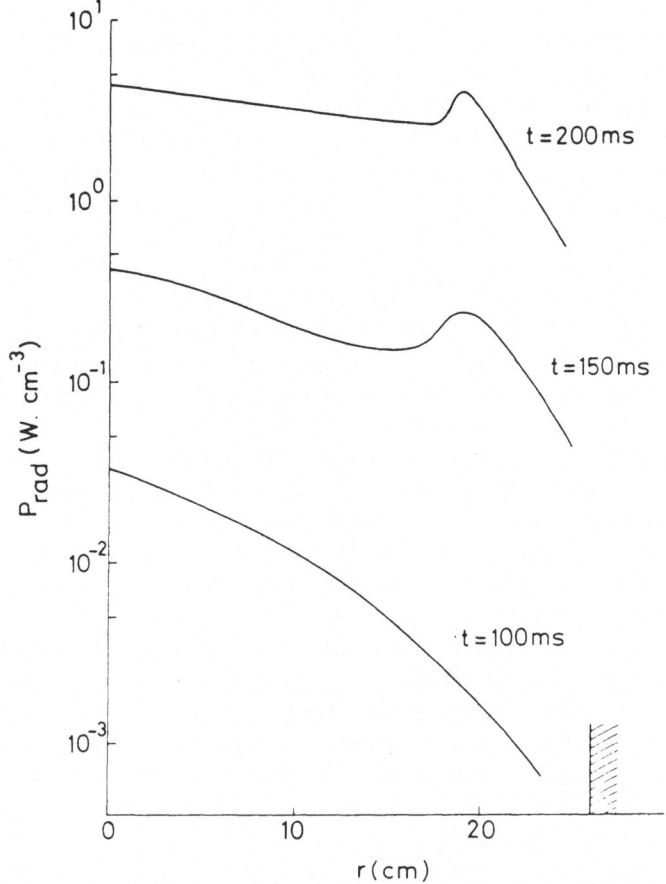

Figure 4.26. Radiated power profiles at various times during DITE discharge D.

culation, and the fact that the radiated power is measured at one location only, can lead to a systematic difference between calculation and experiment.

At the plasma center the radiated power is predominantly from Fe ions and, because in the temperature range of interest the radiated power from Fe ions is relatively insensitive to n_0/n_e, it is possible to estimate that this systematic difference is no more than 10%. At the lower temperatures towards the plasma edge the radiated power from Fe is even less sensitive to n_0/n_e.

In fact at the plasma edge the radiation is, in general, mainly from oxygen ions and the radiated power from these ions is quite sensitive to n_0/n_e. However, at the plasma edge the relative increase in n_0 caused by the injected beams is small due to the much increased values of n_0 compared with the plasma center. This effect reduces the systematic difference between calculation and experiment to an estimated 20%.

While this difference is clearly important in relation to the known experimental uncertainties in P_{rad}, it does not affect our comparison between simulation and experiment in a significant way.

A second question which arises is the sensitivity of the calculated parameters such as temperature and density to the transport coefficients. At high density the electrons and ions have closely coupled temperatures which are then related to all the transport coefficients. However, at lower densities they are decoupled and in a simplified view the calculated electron and ion temperatures are controlled by their respective conductivities and the calculated neutral density is determined by the diffusion coefficient.

An examination of discharge C, for which the effects of radiated power are relatively small, shows that changes of a factor of 1.5 in K_e produce approximately 20% changes in T_e. As T_e is uncertain to $\sim 15\%$ it follows that K_e cannot be certain to better than 40%.

The ion temperature is relatively insensitive to changes in K_i with a factor of 2 change in K_i producing typically up to a 20% change in T_i. The ion conductivity enhancement factor during injection is therefore in the range 5–20 with a range of 2–8 before injection. These figures reflect the experimental uncertainty in T_i (20%) and show the large uncertainty with which the ion thermal conductivity is determined by our experiment.

The diffusion coefficient is constrained by two considerations. If D is increased much above the values used, the diffusion losses become large enough in some cases to extinguish the discharge. Too low a value of D gives a calculated value of n_0 in disagreement with experiment. As n_0 is found to be directly proportional to changes in D, it is concluded that D is determined to about a factor of 2.

Our overall conclusions are:

(i) A wide range of injection heated discharges in DITE can be simulated with a one-dimensional model. Good agreement is found between simulation and experiment although this does not lead to the accurate determination of the transport coefficients.

(ii) The largest uncertainties in the calculations result from inadequate knowledge of the impurity concentrations and from the poorly determined plasma neutral density.

References

E. Agpar, B. Coppi, A. Gondhalekar, H. Helava, D. Komm, F. Martin, B. Montgomery, D. Pappas, R. Parker, and D. Overskei, Plasma Physics and Controlled Nuclear Fusion Research, 1976 (IAEA, Vienna), I, 247 (1977).

P. L. Colestock, S. Davis, P. C. Efthimion, H. P. Eubank, R. J. Goldston, L. R. Grisham, R. J. Hawryluk, J. Hovey, D. L. Jassby, D. W. Johnson, A. A. Mirin, G. Schilling, R. Stooksberry, L. D. Stewart, J. D. Strachan, and H. H. Towner, Princeton Plasma Physics Laboratory Report PPPL-TM-325 (1979).

J. W. Connor, *Plasma Phys.*, **15**, 765 (1973).

J. G. Cordey, K. D. Marx, M. G. McCoy, A. A. Mirin, M. E. Rensink, and J. Killeen, *J. Comput. Phys.*, **28**, 115 (1978).

J. M. Dawson, H. P. Furth, and F. H. Tenney, *Phys. Rev. Lett.*, **26**, 1156 (1971).

H. Eubank, R. J. Goldston, V. Arunasalam, M. Bitter, K. Bol, D. Boyd, N. Bretz, J. P. Bussac, S. Cohen, P. Colestock, S. Davis, D. Dimock, H. Dylla, P. Efthimion, L. Grisham, R. J. Hawryluk. K. W. Hill, E. Hinnov, J. Hosea, H. Hsuan, D. Johnson, G. Martin, S. Medley, E. Meservey, N. Sauthoff, G. Schilling, J. Schivell, G. Schmidt, F. Stauffer, L. Stewart, W. Stodiek, R. Stooksberry, J. Strachan, S. Suckewer, H. Takahashi, G. Tait, M. Ulrickson, S. VonGoeler, and M. Yamada, Plasma Physics and Controlled Nuclear Fusion Research, (IAEA, Vienna) I, 167, (1979).

R. L. Freeman and E. M. Jones, Report CLM-R137, Culham Laboratory Abingdon, Oxfordshire (1974).

H. P. Furth, *Nucl. Fusion* **15**, 487 (1975).

H. P. Furth and D. L. Jassby, *Phys. Rev. Lett.*, **32**, 1176 (1974).

A. H. Futch, J. P. Holdren, J. Killeen, and A. A. Mirin, *Plasma Phys.*, **14**, 211 (1972).

R. D. Gill, Ninth European Conference on Controlled Fusion and Plasma Physics (Oxford, UK), EP9 (1979).

R. D. Gill, G. D. Kerbel, and A. A. Mirin, *Plasma Phys. Cont. Fusion*, **26**, 341 (1984).

F. L. Hinton and R. D. Hazeltine, *Rev. Mod. Phys.*, **48**, 239 (1976).

F. L. Hinton and M. N. Rosenbluth, *Phys. Fluids*, **16**, 836 (1973).

M. H. Hughes and D. E. Post, *J. Comput. Phys.*, **28**, 43 (1978).

J. Hugill, *Nucl. Fusion*, **23**, 331 (1983).

R. A. Hulse, D. E. Post, and D. R. Mikkelson, *J. Phys. B*, **13** (1980), 3895.

D. L. Jassby, *Nucl. Fusion*, **17**, 309 (1977).

D. L. Jassby and H. H. Towner, *Nucl. Fusion*, **16**, 911 (1976).

D. L. Jassby, R. M. Kulsrud, F. W. Perkins, J. Killeen, K. D. Marx, M. G. McCoy, A. A. Mirin, M. E. Rensink, and C. G. Tull, Plasma Physics and Controlled Nuclear Fusion Research, 1976 (IAEA, Vienna) II, 435 (1977).

R. M. Kulsrud and D. L. Jassby, *Nature*, **259**, 541 (1976).

G. G. Lister, D. E. Post, and R. Goldston, Plasma Heating in Torodial Devices (Varenna, Italy) 303 (1976).

K. D. Marx, A. A. Mirin, M. G. McCoy, M. E. Rensink, and J. Killeen, *Nucl. Fusion* **16**, 702 (1976).

A. A. Mirin, A. C. England, H. W. Hendel, D. L. Jassby, and E. B. Nieschmidt, Fokker–Planck analysis of neutron production in beam-injected TFTR plasmas, *Bull. Amer. Phys. Soc.*, **30** (1985).

A. A. Mirin, J. Killeen, K. D. Marx, and M. E. Rensink, *J. Comput. Phys.*, **23**, 23 (1977).

K. Molvig, S. P. Hirshman, and J. C. Whitson, *Phys. Rev. Lett.*, **43**, 582 (1979).

R. E. Olson and A. Salop, *Phys. Rev. A*, **16**, 531 (1977).

J. W. Paul, K. B. Axon, J. Burt, A. D. Craig, S. K. Erents, S. J. Fielding, D. H. J. Goodall, R. D. Gill, R. S. Hemsworth, M. Hobby, J. Hugill, G. M. McCracken, A. Pospiestezyk, B. A. Powell, R. Prentice, G. W. Reid, P. E. Stott, D. D. R. Summers, and C. M. Wilson, Plasma Physics and Controlled Nuclear Fusion Research, 1976 (IAEA, Vienna) II, 269 (1977).

W. Pfeiffer and R. E. Waltz, *Nucl. Fusion*, **19**, 51, (1979).

D. E. Post, R. V. Jensen, C. B. Tarter, W. H. Grasberger, and W. A. Lokke, *Atomic Data and Nuclear Tables* **20**, 397 (1977).

R. D. Richtmyer and K. W. Morton, *Difference Methods for Initial-Value Problems*, 2nd ed., Interscience, New York, 1967.

P. J. Roache, *Computational Fluid Dynamics*, Hermosa, Albuquerque, 1972, p. 161.

D. E. Shumaker, private communication (1979).

A. H. Spano, *Nucl. Fusion*, **15**, 909 (1975).

L. Spitzer, *Physics of Fully Ionized Gases*, Interscience, New York, 1956, p. 79.

J. D. Strachan, P. L. Colestock, S. L. Davis, D. Eames, P. C. Efthimion, H. P. Eubank, R. J. Goldston, L. R. Grisham, R. J. Hawryluk, J. C. Hosea, J. Hovey, D. L. Jassby, D. W. Johnson, A. A. Mirin, G. Schilling, R. Stooksberry, L. D. Stewart and H. H. Towner, *Nucl. Fusion*, **21**, 67 (1981).

D. W. Swain, S. C. Bates, C. E. Bush, R. J. Colchin, W. A. Cooper, J. L. Dunlap, G. R. Dyer, P. H. Edmonds, A. C. England, C. A. Foster, J. T. Hogan, H. C. Howe, R. C. Isler, T. C. Jernigan, H. E. Ketterer, J. Kim, P. W. King, E. A. Lazarus, C. M. Loring, J. F. Lyon, H. C. McCurdy, M. M. Menon, J. T. Mihalczo, S. L. Milora, M. Murakami, A. P. Navarro, R. V. Neidigh, G. H. Neilson, D. R. Overbey, V. K. Pare, Y-K. M. Peng, N. S. Ponte, M. J. Saltmarsh, D. E. Schechter, J. E. Simpkins, W. L. Stirling, C. E. Thomas, C. C. Tsai, J. B. Wilgen, W. R. Wing, R. E. Worsham, and B. Zurro, Ninth European Conference on Controlled Fusion and Plasma Physics (Oxford, UK), B 2.2 (1979).

Index